**Climate
Change
Justice**

Climate Change Justice

Eric A. Posner
David Weisbach

PRINCETON UNIVERSITY PRESS

PRINCETON AND OXFORD

Published by
Princeton University Press,
41 William Street,
Princeton, New Jersey 08540

In the United Kingdom:
Princeton University Press,
6 Oxford Street,
Woodstock, Oxfordshire OX20 1TW

Library of Congress Cataloging-in-Publication Data

Posner, Eric A.
    Climate change justice / Eric A. Posner and David Weisbach.
        p. cm.
    Includes bibliographical references and index.
    ISBN 978-0-691-13775-9 (hardcover : alk. paper) 1. Climate change—
Political aspects. 2. Climate change—Government policy. 3. Climate
change—Law and legislation. I. Weisbach, David, 1963– II. Title.
    QC903.P78 2010
    363.738'74526—dc22

                                                                2009045413

British Library Cataloging-in-Publication Data is available

This book has been composed in Adobe Garamond Pro
with Helvetica Neue display
Printed on acid-free paper. ∞
press.princeton.edu
Printed in the United States of America
10 9 8 7 6 5 4 3 2 1

Contents

Acknowledgments

In preparing this book, we extensively revised some articles that initially appeared in the following places. Eric A. Posner and Cass R. Sunstein, Climate Change Justice, 96 *Georgetown Law Journal* 1565 (2008); Eric A. Posner and Cass R. Sunstein, Should Greenhouse Gas Permits Be Allocated on a Per Capita Basis? 97 *California Law Review* 51 (2009); and Cass Sunstein and David Weisbach, Climate Change and Discounting the Future: A Guide for the Perplexed, 27 *Yale Law and Policy Review* (2010). We thank these journals for permission to republish our articles.

Many people, too numerous to name, have helped us develop the ideas that we advance in this book. Most of them we have thanked in previously published articles. But we should mention David Archer and Ray Pierrehumbert, both of the geophysics department at the University of Chicago. While both David and Ray disagree with much of what we say here, they were kind enough to provide us with advice on the science of climate change as well as criticisms of our arguments. Also thanks to Richard Stewart, Michael Vandenbergh, and anonymous referees for comments on the entire manuscript. We also thank Kathryn Burger, Lena Cohen, and Katie Reaves for research assistance and our dean, Saul Levmore, for financial support.

Cass Sunstein was a coauthor of the three papers mentioned above, and the original plan was for him to coauthor this book as well. Indeed, he was extensively involved in the early stages of the preparation of the manuscript. However, because of his position in the Obama administration—he is currently the Administrator of the White House Office of Information and Regulatory Affairs—he was unable to participate in later revisions. We can only express to him our deepest appreciation for his involvement in this project. He bears no responsibility for any claims in the final version.

Climate
Change
Justice

Introduction

Climate change ranks among the most serious problems facing the world today. There is now a strong scientific consensus that emissions of carbon dioxide and other greenhouse gases into the atmosphere have changed, and will continue to change, the world climate, increasing average temperatures more rapidly than has been seen since long before humans existed. The main source of carbon dioxide emissions is the production and consumption of fossil fuels, but there are many other contributing factors associated with industrial activity and agriculture. In addition, land use changes, including the destruction of forests for farmland, have reduced natural sources of carbon dioxide absorption, further increasing the concentration of carbon dioxide in the atmosphere.

The most optimistic forecast is that climate change will be mild and the changes will happen slowly. Even in this case, local variations in the climate will disrupt traditional economic activities such as agriculture, resulting in the wasting of capital investments, significant migration, and so forth. Even if the sea level rises very little, the dangers from storms will increase, and people will need to build seawalls, to move farther from the coast, and to face other burdens and incur other costs. Warm-weather diseases such as malaria will spread, kill many people, and consequently will need to be seriously addressed.

The median forecast is that under business-as-usual scenarios, global temperature changes will be substantial and the effects of climate change will include severe disruption, with millions of otherwise avoidable deaths caused by flooding, disease, and other hazards, and trillions of dollars in costs. As we shall emphasize, the impacts will probably be worst in the most vulnerable places—poor nations like India and countries in Sub-Saharan Africa. There is also a genuine risk of a truly catastrophic outcome—for example, significant increases in global temperatures and massive sea level rises that would change human life in terrible ways that are difficult to imagine.

What is to be done? A great deal remains unclear, but a few points are straightforward. First, the governments of the leading contributors, now and in the future, should adopt policies that reduce greenhouse gas emissions. These policies will need to make it more costly for people to burn fossil fuels, clear forests, and engage in other activities that contribute to global warming. Voluntary emission reductions by green-minded citizens are a start but are simply not enough. Ideally, governments would set a tax that equals the harm to the climate that results from the use or consumption of a given product. What this means in practice is a complicated issue to which we will return. There are alternative policies such as a cap-and-trade system, where emissions permits are sold and traded. These alternative measures are probably less effective than a tax, but they may be politically more feasible.

Second, whichever policies are chosen—taxes, cap and trade, or something else—governments around the world (or at least all major contributors to the problem) will need to coordinate these policies, most likely through a treaty. The importance of an international treaty can scarcely be exaggerated. Climate change is a global problem. Global warming results from the erosion of a global commons, the capacity of the atmosphere (and the oceans and forests) to store carbon dioxide and other greenhouse gases. If one nation-state heavily regulates emissions within its territory, but no other state does, then the first state will incur heavy costs while producing few benefits for itself or the world. The reason is that emissions continue as before in other countries; those emissions will continue to cause climate change that harms the first state. Worse, industry may migrate from the first state to other states, where it will continue to emit greenhouse gases as before; this is the problem of "leakage." And because of variations in the physical location, geographical features, and level of economic development of different states, it makes good sense to concentrate serious abatement measures in some places—those places where abatement measures are cheaper—rather than others, and to compensate the states that bear the brunt of the abatement costs. This step cannot be undertaken unilaterally. A treaty is necessary.

Many governments have already come round to this view. In the 1990s, nations agreed to the United Nations Framework Convention on Climate Change, which provided some general principles to structure negotiations about a climate treaty. These negotiations yielded the Kyoto Protocol in 1997, which went into force in 2005. The Kyoto Protocol had serious, even fatal problems. It imposed no restrictions on developing nations such as China and India, where emissions are increasing dramatically; for this reason, it could not reduce greenhouse gases to tolerable levels. Further, the Kyoto Protocol imposed an extremely severe burden on the United States, which therefore refused to join the treaty regime. At a climate conference in Bali in December 2008, some progress was made on these issues. The United States committed itself to enter some kind of climate treaty, and developing nations seemed somewhat more willing than in the past to acknowledge that they will have to contribute to greenhouse gas reductions as well.

Meanwhile, diplomats, academics, journalists, and various other commentators have tried to develop principles for the design of the climate treaty. The technical difficulties in designing such a treaty are immense. Nations must agree on the acceptable level of greenhouse gas concentrations in the atmosphere despite enormous scientific and economic uncertainties about the future. They will have to agree on abatement measures and methods for monitoring compliance. And they will have to agree on how the burdens of the abatement measures are distributed. This last question may be the most difficult. Invoking arguments about justice, poor states argue that rich states should bear the bulk of the cost of abatement. China, India, and Brazil urge that the United States and the wealthy nations of Europe should have to undertake aggressive emissions reductions, and should also provide significant financial assistance for any emissions reductions in the developing world. Rich states strenuously object, with some arguing that the biggest emitters (including poor states) should have the strictest obligations and others advancing different principles of equity.

Nations not only argue that different principles of justice apply; they perceive the risks from climate change differently. And even when they agree on the principles of justice, they would apply them differently. Some poor nations perceive themselves as facing grave risks from climate change, and they want rich nations to act immediately to reduce their emissions. Some rich nations believe that they are not gravely threatened by climate change, and they wonder why they should pay a significant cost to solve what is, for them, a speculative problem in the distant future. Some poor nations believe that climate change is far from the most serious threat that they face; they believe that the real threats are connected with poverty, and that economic growth, reducing poverty but increasing greenhouse gas emissions, is highly desirable. Invoking considerations of fairness, they are deeply skeptical about the idea that they should pay a great deal to reduce their own emissions.

In the debate over climate change, questions of fairness and justice play a crucial role. Commentators argue strenuously that a climate treaty should place greater burdens on rich countries than on poor countries, or should punish countries that have contributed the largest amounts of greenhouse gases, or should place equal burdens on people around the world.[1] We contend that the central arguments about justice encounter serious objections. Taken on their own terms—that is, in terms of ideal theory, ignoring pragmatic considerations—they often confuse means and ends. A climate change treaty is not the only method of redistributing wealth and is unlikely to be the best way. If there are better ways of redistributing wealth, we should not use a climate treaty to do so. There are similar problems with other arguments from ideal theory, such as those focusing on corrective justice and past wrongs.

Moreover, even if the arguments were right on their own terms, they fail to consider basic pragmatic or feasibility constraints. By doing so, they threaten to derail a climate change agreement, thus hurting most the nations and people who are pressing those very arguments. There is a tension between ethical claims and pragmatic constraints. To say that nations will refuse to do what they are ethically required to do is not to excuse the violation of ethical requirements. An ethical argument that ignores the interests of states is a fantasy; but a treaty that simply advances the interests of the most powerful states would not have an ethical basis. The challenge is to balance these considerations and construct a treaty that is both ethical and feasible. As we shall see, the two are in some tension, but the tension need not be as destructive as it might seem. We aim to show how a treaty might be feasible, and promote the welfare of people all over the world, while also being consistent with the requirements of justice.

Our argument is unusual. We strongly favor a climate change agreement, especially because it would help poor people in poor nations, and we also favor redistribution from the rich to the poor. At the same time, we reject the claim that certain intuitive ideas about justice should play a major role in the design of a climate agreement. More particularly, we develop four central themes.

Looking Forward, Not Backward

Nations should approach the climate problem from a forward-looking, pragmatic perspective. Many people argue that the climate treaty should reflect principles of corrective and distributive justice. They treat climate negotiations as an opportunity to solve some of the world's most serious problems—the admittedly unfair distribution of wealth across northern and southern countries, the lingering harms of the legacy of colonialism, and so forth. We reject this approach. These arguments run into serious objections of principle. They also threaten to make a climate change agreement far less feasible; any effort to solve all the world's ills, or even some of them, will founder on the diverse values and interests of the various states. Those who make such arguments, including representatives of poor countries but also academics and others in the developed world, are likely in the end to hurt poor countries. If they demand too much from the rich world, the rich world will drag its feet.

Moral Arguments That Make Sense for Individuals Do Not Always Make Sense for States

Governments and commentators who argue about climate change frequently treat states as though they were independent agents that can cause harm and be harmed, that can be culpable, and that have moral obligations to other states. Although this language is an important part of international law and the rhetoric of international relations, it obscures the underlying moral issues. Climate change is a problem because it hurts people, not because it hurts countries. People are often not morally responsible for the harm that results from the policies of states, and care must be taken when moral principles that govern the behavior of individuals—such as principles of corrective and distributive justice—are applied to states.

International Paretianism and the Question of Feasibility

Any treaty must satisfy what we shall call the principle of International Paretianism: all states must believe themselves better off by their lights as a result of the climate treaty.[2] International Paretianism is not an ethical principle but a pragmatic constraint: in the state system, treaties are not possible unless they have the consent of all states, and states only enter treaties that serve their interests. To be sure, states may be influenced by moral arguments, but history supplies very few cases where states act against their own perceived interests in order to satisfy the moral claims of other states. What is true, however, is that states usually define their interests in terms of the well-being of their populations. Thus, the pragmatic constraint of International Paretianism is consistent with a limited but important moral vision—states cooperatively advancing the well-being of their populations, and hence the global population, by agreeing to limits on greenhouse gas emissions. Feasibility rules out the vast redistributions of wealth that many believe are morally required on grounds of corrective and distributive justice, but it does not rule out improvements in global welfare. Feasibility and welfarism are the two pillars of a successful climate treaty.

Globally Optimal Emissions Reductions and the Problem of Local Variation

The optimal climate treaty will provide for a level of greenhouse gas concentrations in the atmosphere that is globally optimal, considering all the effects, good and bad, of emissions reductions. It will set the level of emissions so that the costs of reducing emissions more closely equal the avoided harm, fully taking into account the unknown risks from emissions. Because of variations in the adverse impact of climate change on different countries, the globally optimal level will be higher than what is optimal for some countries, and lower than what is optimal for other countries. It will be necessary to make side payments to the first group of countries in order to secure their cooperation in abatement programs; the second group of countries might have to pay. We suspect that the allocation of burdens will turn in part on the relative bargaining power of the countries, but we believe that ethical considerations will also play a role. A treaty that satisfies International Paretianism will generate a surplus—the climatic benefits minus the costs of abatement—that can be distributed in the form of credits or monetary payments among countries on the basis of ethical postulates. For example, countries that have most aggressively engaged in abatement measures or invested in abatement technologies ought to be rewarded for their efforts. We also agree that wealthy countries should help poor ones with emissions reductions and adaptation, though our claims on this count are qualified. But even if bargaining power ends up determining the division of the surplus, it is important to see that if states are made better off by a treaty, then so will the people living in those states, and that is ethical reason enough for supporting a climate change treaty.

Chapters 1–3 provide the background for our argument. Chapter 1 discusses the scientific and economic facts that bear on the design of a climate treaty, emphasizing an insufficiently appreciated point: the costs and benefits of emissions reductions vary greatly across nations. We show that some nations and regions are far more vulnerable than others; in particular, poor nations are at grave risk, in a way that bears directly on questions of justice. We also emphasize here that a genuinely global treaty is indispensable. Chapter 2 describes the various policy options that are available and considers whether ethical considerations affect the choice of policy options. Chapter 3 describes the local, national, and international efforts that have so far been made and argues that they have been mainly symbolic—no doubt because, however strongly people feel about the problem of climate change, unilateral actions can have little impact on the problem, and so it makes sense to await a treaty rather than put in place expensive but unhelpful regulations.

The core of the book's argument extends from chapter 4 to chapter 7. Chapter 4 criticizes the argument that a climate treaty should reflect the principles of distributive justice—roughly, the view that rich countries should bear the burden of abatement and poor countries either should not have to abate or should be compensated for their abatement efforts. Chapter 5 challenges the argument that the climate treaty should reflect the principles of corrective justice—roughly, the view that countries that industrialized earlier have caused the most harm and should therefore bear the main burden of abatement. Chapter 6 brings these critiques together in order to cast doubt on the popular idea that emissions permits should be distributed on a per capita basis; it also addresses yet another principle—the principle that global resources should be divided equally among the world's inhabitants. Chapter 7 addresses duties to future generations. It insists on a principle of intergenerational neutrality, but offers a qualified endorsement of the view that discounting the future costs and benefits is ethically proper. In chapter 8, we sketch the implications of our arguments for the optimal design of a climate treaty.

At the outset, we should describe some of our ethical and empirical postulates. It is conventional to distinguish between two different approaches to ethical issues: deontological approaches and welfarist approaches. Deontological approaches focus on the rightness or wrongness of particular acts independent of their consequences: for example, certain acts such as lying or harming others are wrong or presumptively wrong, regardless of their consequences. The welfarist approach approves of acts that increase the welfare of relevant people (and possibly animals). For example, an act that increases the welfare of one person without reducing the welfare of anyone else is a good act, and an act that increases the welfare of many people is a very good act.[3]

Our preferred approach to climate change is, as we noted above, broadly welfarist. We find welfarism more congenial and more apt for addressing a phenomenon that is a matter of concern mainly because of its impact on people's well-being. We acknowledge, however, that deontological claims have considerable force, and we shall show that, in many settings, the welfarist and deontological approaches lead to identical conclusions. Finally, we will argue that in certain other settings deontological thinking leads to perverse and intuitively implausible outcomes. It is not, however, our goal to settle the ancient debate between deontologists and welfarists.

A further point is that we will assume, as most welfarists do, certain principles of human equality. We will assume that the welfare of all individuals around the world has equal weight; people in India do not count more or less than people in the United States. We will also assume that people in future generations have equal weight; people born in 2090 do not count more or less than people born in 1990. These assumptions often lead to rather severe prescriptions in favor of the redistribution of wealth: Because poor people would gain more, in welfare terms, from a given unit of money than rich people do, welfarism implies that rich people in rich nations should be transferring a great deal of money to poor people in poor nations. But we also take seriously the state system in which we live, and the practical limits on what ordinary people are willing to sacrifice for the sake of the well-being of others. International Paretianism ensures that we will be discussing only those treaties that have a realistic chance of being ratified.

Chapter 1

Ethically Relevant Facts and Predictions

To study the ethics of climate change, we need to understand how climate change will affect people around the world, now and in the future. That understanding requires knowledge of scientific predictions about the effects of greenhouse gases in various regions of the world and over time, of how people will be affected by these changes, and the extent to which they will adapt. For example, scientific predictions may tell us that a certain region is expected to suffer from a decline in rainfall over the next one hundred years. We need to understand how this change might affect activities in the region, such as agriculture, and the extent to which the people living there will be able to find other sources of food.

It is an understatement to say that there is a vast literature on these issues.[1] Our goal here is to distill this literature into its basic elements so that we have a set of facts and predictions to use in thinking about our ethical obligations with respect to climate change. We rely heavily on the Intergovernmental Panel on Climate Change (the IPCC), the Nobel Prize–winning international agency that was set up to report the scientific consensus on climate-related issues in periodic reports.[2] While there is agreement that climate change is occurring and that it is caused by humans, it is extremely difficult to predict the impact of carbon emissions on people living one hundred or two hundred years in the future. Predictions combine substantial uncertainty on the science of climate change with guesses of how society will evolve over the next several hundred years. Imagine someone living in 1909 or 1809 predicting what life would be like today. We will emphasize here both the central predictions of the effects of climate change and the range of potential outcomes. Many believe that the uncertain possibility of a catastrophe should be the central consideration, and we discuss this as well.

There are five central lessons to be learned from this chapter.

• *Poor nations are likely to suffer most from climate change.* The regions of the world where the climate effects are likely to be the largest are also poor. Poor countries tend to be located in warm regions of the world, and climate change will hurt warm countries more than cool countries. A modest increase in temperature from a relatively cool climate can increase agricultural productivity and reduce the need for heating. A modest increase in temperature from a warm climate, however, reduces agricultural productivity and increases the need for cooling. In addition, the poor are less able to adapt to climate change. Their economies are more dependent on agriculture (which will be hurt by climate change far more than other activities), and they have fewer resources available to mitigate the effects. Nevertheless, most studies of climate change conclude that almost all nations, not just poor countries, are likely to lose from climate change, particularly as the global temperature change rises above minimum amounts.

• *The effects of emissions today will be felt far into the future.* Poor people living in the future, not those living today, are the ones who will be hurt the most. There is a great deal of uncertainty about these effects, because we need to foresee the distant future in order to measure them, and the distant future involves far more uncertainty than anyone can resolve with even a modest degree of accuracy. Uncertainty affects every choice with respect to climate change. How much should we be willing to pay for an "insurance policy" against very bad outcomes? Where, exactly, should we invest resources? In new technologies, in conservation efforts, in ensuring the installation of the best of current technologies? A choice to spend resources today to reduce climate change involves a balancing of the interests of people living today and in the future. We address this issue in detail in chapter 7.

• *People living in poor countries in the future are most likely to benefit from reducing emissions now and in the future.* This conclusion seems unsurprising in light of the fact that poor countries will be most hurt by climate change, but as we will discuss, the issue is more subtle than it first appears. In particular, the data show that if the climate is very sensitive to carbon concentrations, wealthy nations may have the most to gain from reducing emissions. The reason is that if the harm from climate change is sufficiently severe, reducing concentrations somewhat is not sufficient to help poor nations but may make a huge difference for wealthy nations.

• *It is not easy to say how much different countries are responsible for emissions to date because any judgment on that count depends on complex measurement issues.* The standard view, held by the vast majority of analysts, is that wealthy countries are responsible for most emissions to date.[3] The standard view is probably wrong. Developing countries, such as Brazil, Russia, Indonesia, and China, have contributed as much to the stock of greenhouse gases as wealthy countries such as the United States and the countries in the European Union have, at least if we include land use changes and all greenhouse gases as well as carbon dioxide from burning fossil fuels. A similar conclusion follows if we use comprehensive measures of emissions on a per capita basis. With those measures, the top group of emitters includes both developed and developing nations; several small and relatively poor countries, such Belize and Guyana, are at the top of the list; the United States is thirteenth.[4] Even if we limit the list of top per capita emitters to modest and large emitters, the list of top emitters includes a mix of poor and wealthy countries. There are measures under which wealthy countries are primarily responsible for emissions, such as those that include only energy use or that combine countries in various regions. The IPCC, for example, uses a measure that puts responsibility squarely on wealthy countries. Moreover, by any measure, many poor countries, particular those in Africa, emit very little, and almost all rich countries have high or very high emissions.

• *Effective climate action must involve most or all nations with significant emissions.* Any international treaty must include the developed countries who agreed to restrictions on emissions under the Kyoto Protocol (and the United States). But it must also include such countries as China, Brazil, India, and Indonesia. Much of the growth in emissions is in the developing world. Without reductions in emissions in these countries, actions by developed countries will have relatively little effect. As a result, climate treaties that do not include the developing world are unlikely to be effective.

Although some of these conclusions can be disputed, we will take them to be correct for purposes of our discussion. We begin with a very brief overview of the science behind climate change, and then turn to emissions and their likely effects.

The Science

The basic science behind climate change has been understood for more than one hundred years.[5] It is the same science that is used to explain why the Moon is cold, the Earth is the right temperature to support life, and Venus is too hot. The Earth's temperature is determined by the relationship between the energy the Earth absorbs from the Sun and the energy it emits back. These two have to be in balance for the temperature to remain stable. Because the Sun is hot, most of its energy is in the form of visible and near-infrared light. The Earth is much cooler, so the energy it emits back into space has longer wavelengths; it is mostly in the infrared region of the spectrum. Without an atmosphere, the resulting balance of incoming and outgoing energy would mean that the average temperature of the Earth's surface would be about −20°C (which is −4°F)—too cold to support life. The reason the Earth is not this cold is that it is blanketed by the atmosphere. The atmosphere is nearly transparent to incoming solar radiation, but it absorbs the infrared radiation coming from the Earth. It acts like planetary insulation. The effect is to warm the Earth by nearly 35°C, to an average temperature of around 15°C.

The absorption of infrared radiation is due to only minor elements in the atmosphere—the greenhouse gases. The major components of the atmosphere, nitrogen and oxygen, are as transparent to infrared radiation as they are to visible light. The most abundant and significant human-caused greenhouse gas is carbon dioxide.[6] It is only a trace element in the atmosphere, comprising now only about 380 parts per million (ppm), but it has strong effects.

Carbon dioxide occurs naturally in the atmosphere. Pre-industrial concentrations were about 280 ppm.[7] Over the last century or so, however, humans significantly increased the concentration of carbon dioxide, largely through burning fossil fuels, land use change, and agriculture. When we burn fossil fuels, we emit the carbon in these fuels in the form of carbon dioxide. Roughly 60 percent of the annual emissions of greenhouse gases on a global basis come from fossil fuels (and 80 percent of U.S. emissions).[8] Land use change alters the concentration of carbon dioxide in the atmosphere for many reasons. The most important reason is that trees absorb carbon dioxide through photosynthesis. When we cut down trees, we eliminate this carbon sink. In addition, if we burn the timber or it decomposes naturally, we release the carbon that it stored. A little more than 18 percent of global emissions come from forestry practices. Most of the remainder comes from agriculture (13.5 percent globally), which produces emissions from the use of fertilizer and the release of carbon stored in the soil.

Although carbon dioxide is the most important greenhouse gas, a number of other gases have similar greenhouse effects. Many of these have stronger effects, but they are present in lower quantities. Methane is a significant source of the greenhouse effect, estimated to be about twenty-one times as powerful as carbon dioxide. Nitrous oxide is about 310 times as powerful as carbon dioxide. Some gases, such as certain perfluorinated compounds, are almost 24,000 times as powerful as carbon dioxide. These estimates—how much more powerful a gas is than carbon dioxide in terms of the greenhouse effect—are known as "global warming potentials." To measure total emissions, scientists convert emissions of these gases into their equivalent amount of carbon dioxide and add up the total, denominated as CO_2-eq. We will simply refer to carbon dioxide with the understanding that this term includes all gases converted into carbon dioxide equivalents.[9]

All of this would not be a problem except that most greenhouse gases, once emitted into the atmosphere, remain there for a very long time—as far as climate policy is concerned, we can think of them as permanent.[10] Once someone emits carbon dioxide or another greenhouse gas, the effect on the climate is inevitable.[11] In addition, these gases mix evenly in the atmosphere, so the effect is global rather than local. This makes climate change policy different from other pollution policy, where the effects are usually local and often temporary.

Global emissions are now about 50 gigatons of carbon dioxide each year.[12] As a result of these emissions, the concentration of carbon dioxide in the atmosphere (here, just CO_2, not CO_2-eq) has increased from pre-industrial levels of about 280 ppm to about 380 ppm, with average growth rates increasing in the last few years. Currently, concentrations are increasing by about 2 parts per million annually, putting us on a track to double pre-industrial revolution concentrations before 2050 even if emissions stayed constant.[13] If emissions increase with economic growth and if economic growth continues, this may be a floor, and doubling may occur much sooner. Moreover, if other gases are included, concentrations have increased even more, although there is significant uncertainty on the exact numbers when broad measures are used. And we won't run out of fossil fuels when carbon dioxide concentrations double; as long as we keep using fossil fuels, concentrations will increase. If we continue business-as-usual emissions, we can easily quadruple the pre-industrial concentrations.

Average global temperatures have increased over the last century and have been increasing by about 0.2°C each decade for the past thirty years.[14] Figure 1.1 is an example of a commonly seen graph of global surface temperatures.[15]

A standard question is whether these temperature increases are due to natural causes rather than human-caused emissions of greenhouse gases. After all, temperatures have fluctuated dramatically in the past, causing warm periods and ice ages. And our understanding of how aerosols, clouds, and other natural phenomena affect the climate is uncertain. How can we be sure this is not natural variation?

Scientists have tried to determine the extent of human causation as opposed to natural variability in a number of ways. The central argument is that the temperature changes over the last century can only be explained by models that include estimates of both natural

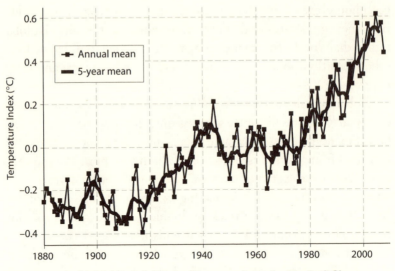

Figure 1.1 Global Land-Ocean Temperature Anomaly (°C).

variability and human-caused climate change. If there is no human-caused climate change, the models cannot explain the temperature record; in fact, they show that the likely natural causes of climate variability over recent periods would have produced a slight cooling rather than warming.[16] Moreover, an alternative explanation would have to address why the basic science, that carbon dioxide produces a greenhouse effect, is wrong. As David Archer, a climate scientist at the University of Chicago, observes, to have an alternative explanation of the temperature record, we would need not one but two entirely novel scientific discoveries: one to explain what is causing the temperature increase and another to explain why the carbon dioxide is not trapping heat as expected.[17] By analogy, if we find the butler standing in front of a victim with smoke coming out of his gun, it is possible that someone else is the criminal, but we would have to be able to find that someone else and explain the gun and the smoke. This is a tall order. The IPCC therefore concludes that "it is *extremely unlikely* that global climate change of the past 50 years can be explained without external forcing [i.e., emissions of greenhouse gases]."[18] We will take significant human-caused climate change as entirely beyond debate.

All of what we have just reported is accepted by nearly every climate scientist, although some of the data, such as the extent of emissions of non-CO_2 gases and some temperature data, are subject to measurement uncertainty. The hard problems in climate science involve nailing down the precise effects. A central variable is what is known as *climate sensitivity*. This term refers to the equilibrium global average surface temperature increase for a doubling of CO_2. One reason this is hard to predict is that temperature changes happen slowly. There is, for example, significant temperature inertia in large masses, such as the oceans. We cannot observe immediate effects of actions, which means that observations have to be combined with models to predict equilibrium changes. Moreover, the climate system is very complex. The simple description of the causes of climate change given above ignored such factors as humidity, convection, wind, and aerosols. Feedback effects, such as changes in how much light is reflected back into the sky due to melting ice, are difficult to calculate. Cloud feedbacks are particularly difficult to predict and are the single largest source of uncertainty.[19]

Given these uncertainties, the most recent IPCC estimate for climate sensitivity is 3°C, with a likely range of between 2°C and 4.5°C. This is a broad range; the differences in effects on human beings and the environment between 2°C and 4.5°C would likely be dramatic. Moreover, this range does not represent the full spread on the high end since the IPCC does not fully incorporate a variety of feedback mechanisms, such as cloud feedbacks and soil carbon release. Worse, this range does not reflect the full level of uncertainty because a key variable is how fast the changes occur. The effects of climate change will depend critically on speed: slow climate change gives humans and other species time to adapt, while rapid climate change does not.

To get a better sense of the range of uncertainty about climate sensitivity, the UK government, in a major study of the economics of climate change, known as the Stern Review, collected a range of estimates from various models the likelihood of a given global average temperature change for given carbon concentrations.[20] In table 1.1, we reproduce the estimates for the Hadley Centre Ensemble of models.[21] The numbers are highly uncertain and are best viewed as a

Table 1.1.
Likelihood (in percentage) of Exceeding a Temperature Increase at Equilibrium

Stabilization Level (ppm of CO_2e)	2°	3°	4°	5°	6°	7°
450	78	18	3	1	0	0
500	96	44	11	3	1	0
550 (doubling)	99	69	24	7	2	1
650	100	94	58	24	9	4
750	100	99	82	47	22	9

rough guide. The table indicates that there is a 99 percent chance of a 2°C temperature increase for a doubling of CO_2. Moreover, there is a 24 percent chance of a 4°C increase, which would be very harmful, and even a 1 percent chance of a 7°C increase, which would be truly catastrophic. As concentrations of CO_2 rise, the chances of very large temperature increases correspondingly increase. Because of the risk of catastrophe, good arguments exist that the risks of large temperature increases, not the expected increase, should drive policy.[22]

Finally, as noted, the timing of these changes is difficult to predict, but we know that the changes will occur in the future, possibly in one hundred or two hundred years. The climate system has significant inertia so that emissions today will continue to cause the climate to change over a long period of time. Even more important, carbon dioxide concentrations continue to increase because of new emissions. Our economy and its reliance on fossil fuel energy and modern forms of agriculture cannot be changed overnight. Future emissions are inevitable, and thus future temperature increases are inevitable. Given these factors, models predict temperature increases over long periods of time.

Impacts

All of this would be of little interest as a matter of social policy if it didn't have impacts on humans and other living creatures. The central question, therefore, is what are the expected effects of climate change? What does business as usual mean for people living around the world?

The first step in estimating the effects is determining what business-as-usual will be. This is no small task; it involves predicting emissions long into the future, which means predicting population changes, economic growth, and technologies far into the future. Making such predictions has been done through "emissions scenarios." Scenarios are future histories of the world, storylines that present possible ways that the next one hundred years will unfold. The IPCC put together a series of these, which represent a range of possibilities, such as a convergence story in which the developing world rapidly joins the developed world, a divergence story, a regional development story, and so on.[23] Each of these stories has implications for economic and population growth and, as a result, for emissions.

If we can guess future emissions through the use of scenarios, we can then translate this into temperature increases through measures such as climate sensitivity. This, however, only gives us the global average surface temperature as a measure of the effect of carbon emissions. Temperature changes will not be uniform, which means that we need to translate global averages into local and seasonal effects. These effects will vary widely. For example, northern latitudes are expected to face greater warming than those around the equator.

Even disaggregating the world into regions does not reflect the true effects of climate change. Local averages are made up of seasonal weather patterns, cloudy and clear, hot and cold, wet and dry. At a given location, such as the southwestern United States or central China, it is misleading to think of climate change in terms of a single number, temperature change. Instead, we should think of temperature change as merely an index of the overall disruption in local weather patterns. Small changes in the index can lead to large local disruptions. We might, for example, see disruption of monsoon patterns in some parts of the world, with the resulting harms to agriculture, increased rain elsewhere, changed growing seasons, sea level rise, and any number of other local effects.

Local effects of this sort are extremely difficult to predict, and scientists are not yet at the point where they feel confident making these predictions. Nevertheless, because of its importance, there is a large body of work estimating the likely effects in regions around the world—all of the IPCC Working Group II is devoted to this, and their Fourth Assessment Report is just under one thousand pages of dense prose.[24] In the report, they detail local effects such as the loss of the Himalayan glaciers, which provide water to hundreds of millions of people. The Gangetic basin alone is home to 500 million people, who will all face severe water shortages with the loss of these glaciers. Similarly, the IPCC details the reduction in water availability and a shortened growing season in parts of Africa that already face significant food and water shortages. In the United States, water resources that are already over-allocated will face additional pressure, agricultural patterns may have to change, and sea level rise may threaten coastal communities. Southern Europe is likely to face significant droughts, resulting in lower agricultural productivity and increased risk of fires. Bangladesh may lose a substantial fraction of its land to the sea, resulting in a refugee crisis in the region.

There is a risk of getting lost in detail; what would be most useful is a set of take-home facts that reflect the central, although uncertain, expectations of what is to come absent emissions reductions. We emphasize three.

The first is that there will be an overall loss in global welfare. The loss increases with temperature increases and, at even modest levels, may be more than the costs of reducing emissions. Full cost-benefit analyses of climate change thus far have been limited in large part because of the complexity of the problem; the best analyses have been able to do is estimate market impacts of climate change and make best guesses about the costs of reducing emissions. They do not know what it will cost to reduce emissions, and examining only market impacts leaves out many of the most important impacts, such as those on migration and national security. Nevertheless, global emissions reductions at a fairly stringent level likely pass a cost-benefit test. This is important because it means that reducing emissions should lead to an overall increase in welfare; even if there are some nations or regions that are worse off because their costs of emissions reductions exceed their benefits, globally this will not be the case, and these nations, if necessary, can be compensated. This is true even though many of the costs of abatement will be incurred today and the benefits realized in the future: reducing emissions passes a cost-benefit test on a present value basis.

The extent of the necessary reductions and the speed of the reductions are subject to significant disagreement. The problem is enormously complex, requiring modelers to estimate parameters such as local climate conditions, impacts on those living in local regions, possible adaptations, and technological progress in producing low carbon energy. These all have to be combined in a global general equilibrium model which estimates conditions for the indefinite future. Given these complexities, modelers have taken a variety of approaches. Central modeling assumptions, such as the cost of adaptation and the rate of technological progress, vary; different functional forms are used to represent economies; and basic structures of the models, such as how many regions they use, how they model uncertainty, and how they estimate damages from climate change (and even what types of damages are modeled) are not consistent. It is, therefore, not easy to compare one model to another and to understand the reasons for disagreement.[25]

Given the current state of knowledge, we do not take any position here on the extent of the necessary reductions. It appears as if almost all studies to date indicate that at least a modest reduction in emissions is cost beneficial, and a number of studies indicate that fast and deep emissions reductions are needed. Moreover, all of the existing studies, even those that are only a few years old, are now out of date; emissions have been rising faster than anticipated and the climate effects appear to be more severe than the central estimates. Unfortunately, we may need fast and deep cuts in emissions.

Second, poor countries are likely to be hurt far more than rich countries. As we mentioned, one reason is bad luck: the regions of the world where the effects of emissions will be the worst also happen to be poor, and poor countries tend to be located in warm regions of the world, where the effects of these changes will be entirely negative. In addition, poor countries tend to be more dependent on agriculture than rich countries, which means that they are more vulnerable to any given level of change. Finally, poor countries cannot adapt as easily as rich countries, simply because of lack of resources. Africa in particular appears to be the most vulnerable, although small islands and low-lying areas outside of Africa (such as Bangladesh) are also significantly at risk.[26] South and Southeast Asia (especially India) are also at significant risk. OECD Europe may face serious problems, although it has a much greater ability to adapt than countries in Africa and Asia.

The Center for International Earth Science Information Network (CIESIN) at Columbia University produced a set of graphical estimates of impacts by country.[27] The Center calculated vulnerabilities to climate change by country based on preexisting indices of vulnerabilities to environmental stresses. The Center's measure includes both sensitivity to climate stresses and adaptability. Under the resulting index, the least vulnerable countries in the world are the Scandinavian countries, Switzerland, Austria, France, Belgium, Italy, Japan, Canada, and the United States. China is right in the middle, exhibiting about average vulnerability. Of the fifteen most vulnerable countries, fourteen are in Africa, with Bangladesh the only country outside of that continent. Non-African countries in the bottom third are located largely in Central or South America.[28] The authors combined this data with a climate model that gives regional impacts. This model allowed them to estimate vulnerability by country. The exact effect depends on future scenarios (i.e., what are future emissions, how sensitive is the climate to emissions, how fast do various economies around the world grow, etc.). They produced maps showing various possibilities.

We reproduce two of their maps here. The first, figure 1.2, is a measure of vulnerability in the year 2100 under one of the IPCC scenarios (a scenario that they call A2) and where the climate turns out to be quite insensitive to carbon. As can be seen, many countries have some vulnerability—but the worst problems are in Africa, followed by central and eastern Asia, central and southern South America, and parts of the Middle East and Eastern Europe.

The second map, figure 1.3, is the same scenario but with a high climate sensitivity. As can be seen, vulnerabilities are much worse.

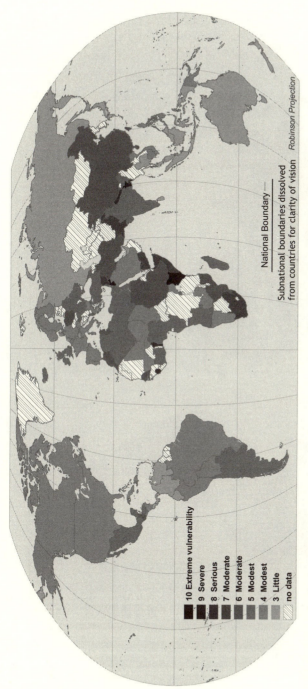

Figure 1.2 Global Distribution of Vulnerability to Climate Change. Combined National Indices of Exposure and Sensitivity. Scenario A2 in year 2100 with climate sensitivity equal to 1.5°C annual mean temperature with aggregate impacts calibration.

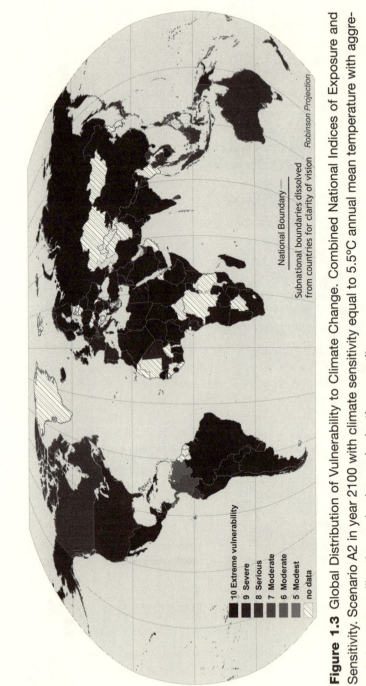

Figure 1.3 Global Distribution of Vulnerability to Climate Change. Combined National Indices of Exposure and Sensitivity. Scenario A2 in year 2100 with climate sensitivity equal to 5.5°C annual mean temperature with aggregate impacts calibration and enhanced adaptive capacity.

Table 1.2.

Damages of a 2.5°C Warming as a
Percentage of GDP

India	4.93
Africa	3.91
OECD Europe	2.83
High-income OPEC	1.95
Eastern Europe	0.71
Japan	0.50
United States	0.45
China	0.22
Russia	–0.65

Only a very few countries, including the United States, are moderately affected, with virtually all of Africa and Asia showing extreme vulnerability. One way of interpreting the two maps is that they illustrate both expectations (which countries are most likely to be hurt) and risks (how damages change with climate sensitivity). Some countries are extremely likely to be hurt regardless of climate sensitivity, while other countries have more variable outcomes.

Very few analysts are willing to venture estimates of the changes in GDP from climate change in various regions of the world. The exercise is simply too speculative. As we noted, the IPCC, for example, reports only disaggregated data on particular harms. In an influential book,[29] William Nordhaus and Joseph Boyer give an estimate, which we report here as a crude but potentially informative guess. Their numbers are presented in table 1.2.

The numbers reflect somewhat different estimates from the Columbia study. For example, the Columbia study puts China somewhere near the middle in terms of climate vulnerability, while Nordhaus and Boyer claim that it is less vulnerable. The results for India and Africa, however, are similar.

Our third and final take-home fact about impacts is that there is a huge amount of uncertainty about the expected effects of climate change. This should be evident from the previous discussion, but it has independent importance. Even if we thought that the central estimates of the impact of climate change were acceptable costs for having cheap energy, for example, we might still want to reduce emissions as insurance against the possibility of very bad outcomes. The uncertainty about the likely impacts of climate change is skewed: the best-case scenarios (the climate is relatively insensitive to carbon dioxide, the changes happen slowly, and adaptation is less expensive than anticipated) will mean that there are only modest costs, but the worst-case scenarios are very bad. The last time the temperature was six degrees warmer than today was somewhere around 55 million years ago, long before humans existed. To get a sense of what six degrees means, the last great ice age, in which substantial parts of North American and Europe were covered by ice sheets, had global average surface temperatures that were six degrees cooler than today—six degrees difference in temperatures can produce a very different world.[30]

This means that in the unfortunate event that outcomes are worse than expected, they may be very much worse, and all nations will suffer. That is, even nations that might estimate that they do not expect to suffer severe consequences from climate change must also calculate a risk that they might, and these nations might be willing to enter into a climate treaty as an insurance policy in the event of a catastrophe.

Uncertainty also obscures the regional distribution of harms. The data above gave the expected harms by country or region, but various countries or regions around the world will have different attitudes toward risk. They will therefore evaluate the overall set of payoffs differently. If poor nations are less tolerant of risk than wealthy nations, uncertainty is likely to magnify the disparities illustrated above.

Benefits from Abatement

As a general matter, the same nations that are likely to be most hurt by climate change are the mostly likely to benefit from abatement. Because climate change is expected to hurt the poor the most, climate change abatement will also help the poor. Because most of the harms from climate change will be in the future, emissions abatement should be thought of as helping the future poor. Reducing carbon emissions is not a way to help today's poor. In fact, the opposite is true: to the extent that resources are devoted to reducing carbon emissions, these resources are not available to help today's poor.

Even though on average emissions abatement is expected to help future poor people, the analysis is actually more complicated than that. In scenarios in which the damage from climate change is very severe, developed countries, including the United States, will benefit most from abatement. The reason is that if harms from climate change are very severe, rich nations will face an impact akin to that expected for poor nations. Abatement may make it possible for richer nations to mitigate the harm while not significantly helping poorer nations. To illustrate, consider figure 1.3 above. This figure shows the vulnerability to harm from climate change in the year 2100 under the same scenario used previously (A2) with the assumption that the climate is very sensitive to carbon dioxide concentrations (sensitivity of 5.5°C).

The authors of that study then consider the effects of limiting concentrations to 550 ppm. The results are shown in figure 1.4. The countries that are helped are those whose color changes to lighter shades. As can be seen, in this scenario, it is largely the wealthier countries that benefit. Much of Africa, for example, remains severely harmed. One way to understand this effect is to imagine a set of balls lined up in a vertical tube, with water at some level below the balls. If we push the balls down the tube just a bit, the balls at the bottom are the first to go into the water, and are also the ones to rise above the water if we let them rise. If we push down a lot, however, the first ones to rise above the water if we let up a bit are the ones that reach it later. A little bit of help is not sufficient for the balls that are at or near the bottom. Similarly, if the effects of climate change are likely to be severe, remedial action will benefit the nations that are moderately hurt more than the nations that are severely hurt. This result does not hold if climate change results in less severe harm.

We should also emphasize that poor nations that are likely to be hurt by climate change are not necessarily *best* helped by climate change abatement. They may be better off if the resources that could

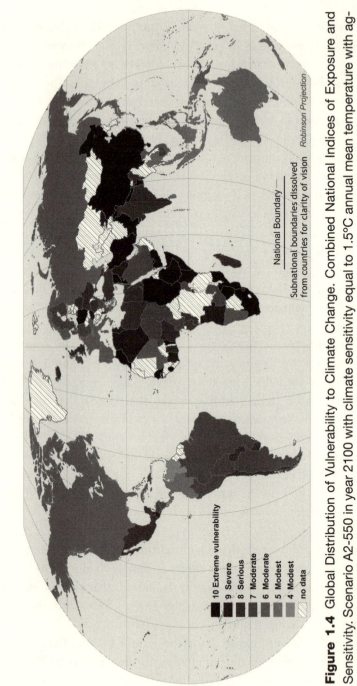

Figure 1.4 Global Distribution of Vulnerability to Climate Change. Combined National Indices of Exposure and Sensitivity. Scenario A2-550 in year 2100 with climate sensitivity equal to 1.5°C annual mean temperature with aggregate impacts calibration.

10 Extreme vulnerability
9 Severe
8 Serious
7 Moderate
6 Moderate
5 Modest
4 Modest
no data

National Boundary —
Subnational boundaries dissolved
from countries for clarity of vision   *Robinson Projection*

be dedicated toward abatement are used in other ways. Consider the case of malaria.[31] Increased temperatures have the potential to increase the incidence of malaria, which occurs only in warm climates. If carbon emissions do not decline, malaria ranges are predicted to spread into the southeastern United States as well as much of China by 2050.[32] Malaria, however, is a disease of the poor. According to the World Health Organization, those with an annual income of $3,000 or more do not die of malaria. In the IPCC scenarios used to predict the effects of climate change, income in all regions of the world will be growing. According to one standard scenario used by the IPCC, all regions of the world will have sufficient per capita income to surpass the $3,000 threshold by 2085. Under this scenario, therefore, deaths from malaria (although not other harms from the disease) will be eliminated in 2085. If resources are diverted to reduce climate change, incomes will be systematically lower, delaying the time when all regions surpass this threshold.

It follows that there is a trade-off between reducing malaria by reducing temperature changes and reducing malaria deaths by increasing incomes. On the basis of this reasoning and a computer model of the world economy, two authors predict that malaria deaths are minimized with relatively small reductions in emissions.[33] In fact, the Kyoto Protocol could cause an *increase* in malaria deaths—if it slows development and hence prevents people from reaching the $3,000 level while having no effect on climate change. If one merely looked at the predictions of the spread of malaria due to climate change, one would get a very misleading picture of the problem. Malaria may be more complex than these models. Scientists do not know with certainty whether using resources to reduce emissions will increase or reduce its incidence, but the basic point remains: so long as resources are limited, we face complex trade-offs.

International Agreement Necessary

To achieve climate change abatement, all major emitting nations, including developing nations, will have to reduce their emissions. As we will discuss in chapter 3, the Framework Convention on Climate

Change envisioned "common but differentiated responsibilities."[34] The Kyoto Protocol, however, left out developing nations.[35] This was a mistake. We will explore the reasons for favoring developing nations later, but it is plain that all nations need to reduce emissions.

We can see this through simple back-of-the-envelope calculations. The current carbon dioxide concentration is about 380 ppm and is growing at roughly 2 ppm per year. If this rate stays the same, the carbon dioxide concentration would reach about 460 ppm by mid-century. The world economy, however, is expected to be around three times larger by mid-century than it is today. If the economy is three times larger, emissions will have to be three times lower per unit of GDP, just to keep concentrations at 460 ppm by 2050, not to stabilize them. Agriculture and land use together make up about 34 percent of emissions and these emissions, particularly those from agriculture, are likely to be difficult to cut. If emissions reductions come from the energy sector and not agriculture and land use, however, we would need to be able to produce sufficient energy to support the world at three times current GDP with zero emissions. Even if land use emissions can be cut or if agriculture and land use emissions do not grow with GDP (unlike energy), we would still need close to zero emissions from energy to achieve this goal. Stabilization of concentrations would be even more difficult, almost certainly requiring zero or close to zero emissions from energy and a complete stop to deforestation.[36]

The developed countries cannot achieve these goals by themselves. China is now the world's largest emitter. After the United States, which is second, Indonesia, Brazil, Russia, and India are next. Without deep cuts by these countries *from current levels*, it is impossible to achieve reasonable stabilization goals.

Moreover, much of the predicted growth in emissions over the next few decades is expected to come from developing countries. Consider figures 1.5 and 1.6. Figure 1.5 is the International Energy Agency projection for emissions (from energy) for the United States from 2006 until 2030. It shows an increase in emissions over the next twenty years.[37]

Figure 1.6 shows the same projection for the United States, but also includes the rest of the world, broken into six additional

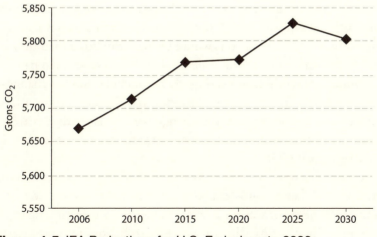

Figure 1.5 IEA Projections for U.S. Emissions to 2030.

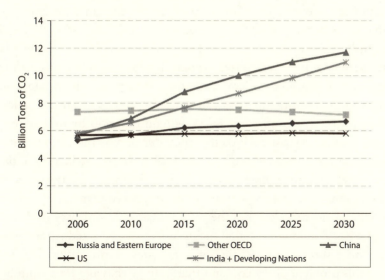

Figure 1.6 IEA Global Emissions Projections to 2030.

regions. As can be seen, the increases in U.S. emissions are swamped by increases elsewhere—the U.S. projections look flat. Moreover, even if the United States and the other OECD countries reduced their emissions dramatically, the increases in the rest of the world would swamp these reductions. The *increases* in emissions in China,

Russia, India, and other non-OECD countries are larger than the *total* OECD emissions in 2006; even if OECD countries had zero emissions in 2030, emissions would be higher in 2030 than they are today if other countries continue on their business-as-usual path.

Note also that these projections do not include land use change—they are International Energy Agency projections of emission from energy use. Including land use change would make the chart even more skewed because almost all emissions from land use change are outside the OECD.

Even if this were not true—say that emissions in developing countries were at levels consistent with stabilization goals and were not growing rapidly—it would be important to include developing countries in a climate treaty. The reason is that reductions in energy use by only a few countries will simply lower the price of energy for other countries, allowing them to consume more. Imagine, for example, that oil exporting nations continue to pump oil as fast as they can regardless of the world price (at least within reason). This is roughly consistent with experience with oil production and oil prices over the last several years—as prices have changed dramatically, production has not significantly increased. In these circumstances, any reduction by one country is offset by increased consumption by other countries.[38] In effect, costly efforts to reduce demand by the United States act simply as a transfer to other countries through a reduction in the price at which they must purchase energy. With certain assumptions about how oil exporting nations react to reduced demand, the effect might be more modest. But the point remains: the only way to reduce the use of fossil fuels is for all major users to agree to reductions.

To be sure, many people argue for conservation notwithstanding this problem. They contend that only by showing leadership on the issue can developed countries convince developing countries to agree to conversation measures.[39] This may be true, but it is essential to keep in mind that the goal of any action must, in the end, be to reduce the worldwide supply of fossil fuels, not the demand for fossil fuels in a few countries.

Finally, we will discuss how much the United States and other rich nations should pay relative to other nations for abating carbon emissions. Regardless of the answer to this question, we should not attempt to concentrate abatement in rich countries. Even if rich nations should pay most of the cost of reducing emissions, the decisions about where around the globe to abate should be based on finding the lowest-cost opportunities to reduce emissions. The reason is straightforward: reducing carbon emissions and preventing the harms from climate change is likely to be expensive. It is important, in order to minimize these costs, to identify the lowest cost abatement opportunities regardless of where they are.[40] Imposing additional abatement requirements on one nation or set of nations while exempting others risks significantly increasing the costs, to the detriment of all.

Economists have estimated the cost of pursuing climate change using a subset of countries. One study, for example, considered the costs of meeting the Kyoto targets with and without trading across countries, in various permutations (i.e., no trading, only trading within Annex I, trading across all countries, etc.).[41] The costs decrease dramatically as more nations are included in the trading regime, dropping by more than 93 percent from the "no-trading" case to the "trading across all countries" case. When considering the size of the global restructuring needed to reduce carbon emissions, these savings are large indeed. Other studies have found similar results.[42]

A central question of justice will be how to pay for emissions reductions—we address this in detail throughout the rest of the book—but as a simple matter of mathematics, the world cannot achieve stabilization at reasonable levels, such as 450–500 ppm, without substantial cuts in emissions by all nations. There may be some headroom in which developing countries can increase emissions in the very short run, but all nations will have to reduce emissions soon.

Emissions and Emitters

We close with a look backward at who has emitted in the past. Under some theories of justice, this matters. The standard view is that wealthy nations are the source of most greenhouse gas emissions. According to the IPCC, "industrialized nations are the source of most past and current GHG emissions."[43] The data, however, tell a more

Table 1.3.

Greenhouse Gas Emissions in 2000

CO_2, CH_4, N_2O, PFC's, HFC's, SF_6, and Land Use Change

Rank	Country	$MtCO_2$	% of Total	Tons Per Person	Rank	$ Per Person
1	United States of America	6,443	15.5	22.8	13	36,451
2	China	4,771	11.5	3.8	120	5,490
3	Indonesia	3,066	7.4	14.9	24	3,282
4	Brazil	2,314	5.6	13.3	33	7,406
5	Russian Federation	1,960	4.7	13.4	32	9,021
6	India	1,553	3.7	1.5	168	2,851
7	Japan	1,317	3.2	10.4	51	27,114
8	Germany	1,006	2.4	12.2	37	25,945
9	Malaysia	852	2.0	36.6	5	9,374
10	Canada	766	1.8	24.9	12	29,136

complicated story, and under many, if not most, good measures of emissions, a broad range of countries, including both developed and developing countries, share responsibility for most past emissions. Moreover, under almost any measure, much of the growth will take place in developing countries.[44]

We begin with a broad measure of annual emissions including the six most important greenhouse gases and emissions from land-use change and forestry. The most recent data using this measure of emission are from 2000, measuring emissions in millions of metric tons of CO_2 equivalents ($MtCO_2$e). Table 1.3 lists the top ten emitters in 2000.

The World Bank defines high-income countries as countries having per capita income of more than around $11,500.[45] Under this standard, the wealthy countries on the top ten list (the United States, Japan, Germany, and Canada)[46] are responsible for about 23 percent of current emissions and others are responsible for 35 percent. The analysis does not change if we expand the list: the top forty countries

Table 1.4.

Share of Global Carbon Dioxide Emissions (not including land use), 2005

CO_2, CH_4, N_2O, PFC's, HFC's, SF_6 without Land Use Change

Rank	Country	$MtCO_2$	% of World Total	Tons CO_2 Per Person	Rank
1	China	7,219	19.1	5.5	71
2	United States of America	6,964	18.4	28.5	7
3	Russian Federation	1,960	5.2	14	18
4	India	1,853	4.9	1.7	119
5	Japan	1,343	3.6	10.5	37
6	Brazil	1,014	2.7	5.4	73
7	Germany	977	2.6	11.9	25
8	Canada	732	1.9	22.6	8
9	United Kingdom	640	1.7	10.6	36
10	Mexico	630	1.7	6.1	64

ranked by GDP per capita emit about 34 percent of greenhouse gases under a broad measure of emissions. Moreover, although these data are fairly recent, they do not reflect the rapid changes in emissions from China and other fast-growing developing nations. China became the top emitter as of 2005, emitting 7.2 billion tons of carbon dioxide or equivalents (excluding land use change), compared to U.S. emissions of 6.9 billion in 2005. Russia's emissions remained relatively constant between 2000 and 2005, but India's grew to 1.9 billion tons, putting it in a virtual tie with Russia as the third largest emitter in the world.

Most discussions of emissions focus only on carbon dioxide emissions from energy use. Based on this measure, the list of top emitters changes. Table 1.4 shows the 2005 list when land use is excluded. Narrowing the measurement by excluding land use change removes many less wealthy or relatively poor countries such as Brazil, Indonesia, and Malaysia from the list of top emitters. But it is hard to

Table 1.5.

Cumulative Emissions: CO_2 from Energy and Land Use Change, 1950–2000

Rank	Country	$MtCO_2$	% of World Total	Tons CO_2 Per Person	Rank
1	United States of America	184,827	16.9	623.3	11
2	China	108,117	9.9	82.9	111
3	Russian Federation	90,068	8.3	629.2	10
4	Indonesia	79,996	7.3	362.7	35
5	Brazil	68,029	6.2	364.1	34
6	Germany	46,382	4.3	562.4	13
7	Japan	41,603	3.8	325.6	41
8	United Kingdom	29,164	2.7	484.2	18
9	Canada	22,642	2.1	700.7	7
10	Malaysia	22,228	2.0	866.5	4

see any justification for using this narrow measure; an emission of a ton of carbon dioxide or its equivalent affects the climate the same regardless of its source.

The numbers given above refer to *flows*: how much a given nation emits on an annual basis. Climate change, however, is caused by the *stock* of emissions: the total concentration of carbon dioxide and other greenhouse gases in the atmosphere. Because most greenhouse gases have very long lifetimes in the atmosphere, we have to look at emissions over time to determine contributions to the stock of greenhouse gases.[47] Once again, broad measures of emissions, including land use change, are available only for 2000. In addition, broad measures only go back to 1950 and do not include greenhouse gases other than carbon dioxide. Table 1.5 presents the data.

The group of wealthy countries on this list, the United States, Germany, Japan, the United Kingdom, and Canada, makes up about 29 percent of cumulative emissions.[48] The poor countries in this list make up about 33 percent of cumulative emissions. The basic result for stocks of emissions is the same as for flows: both wealthy and poor countries are major contributors. If we narrowed our focus to

Table 1.6.
Cumulative Emissions of CO_2 (including land use change)
Per Capita, 1950–2000

Rank	Country	Per Capita Emissions (stock)
1	Belize	3,390
2	Guyana	2,146
3	Luxembourg	1,310
4	Malaysia	900
5	Papua New Guinea	759
6	Panama	709
7	Canada	708
8	Czech Republic	665
9	Estonia	664
10	The United States	636
11	Russia	634
. . .		
20	UK	497
33	EU-25	385
111	China	85
165	India	16

exclude land use changes, the wealthy countries would look like a larger source of emissions, just as with flows.

There are a variety of alternative measures of relative emissions. The two most common additional measures are emissions per capita and emissions intensity (i.e., the emissions from a given country necessary to produce a dollar of output). Table 1.6 lists the top ten nations by per capita contributions to stocks of emissions.[49]

The standard view that wealthy countries are responsible for most emissions often uses per capita data, but the data are aggregated across large groups of countries. For example, in the year 2000, the top twenty countries in terms of income had about one-eighth of the world population (800 million people out of 6.3 billion total world population), but emitted about 40 percent of the total emissions (all gases including land use change). If we look only at carbon

Table 1.7.

Greenhouse Gas Intensity of Economy, 2000 (CO_2, CH_4, N_2O, PFCs, HFCs, SF_6, land use change)

Rank	Country	$tCO_2\text{-}e$/Million $	Index
1	Zambia	31,292	100
2	Belize	16,524	52.6
3	Liberia	14,932	47.5
4	Congo, Dem. Republic	13,560	43.1
5	Guyana	13,390	42.5
6	Papua New Guinea	12,491	39.7
7	Sierra Leone	8,998	28.5
8	Myanmar	8,764	27.7
9	Central African Republic	7,487	23.6
10	Mongolia	7,420	23.4
...			
98	Australia	1,018	2.8
100	China	972	2.7
123	South Korea	693	1.8
126	United States	662	1.7
158	EU 25	455	1

dioxide from burning fossil fuels and aggregate back to 1850, these same countries are responsible for about 58 percent of the total stock of carbon dioxide. The bottom billion people (grouped by country GDP per capita, not the actual individual incomes) emitted less than 2 percent of the stock of carbon dioxide from energy use. The difference in results arises because of the narrow focus on emissions from fossil fuels (which correlates highly with wealth) and because of aggregation: poor countries that are very high per capita emitters go into the very large pool of poor countries, so their high emissions get averaged with poor countries that are low emitters.[50]

A final way that emissions are commonly measured is by intensity. Table 1.7 lists the top ten countries by the number of carbon or other greenhouse gases emitted (measured in equivalent tons of carbon) per million dollars of GDP, measured using purchasing power parity

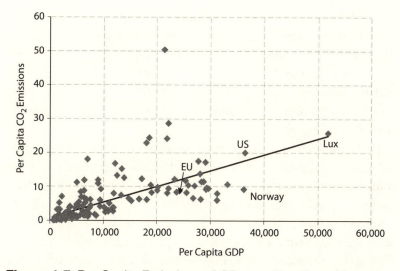

Figure 1.7 Per Capita Emissions of CO_2 as a Function of Income.

adjustments. Many of these countries are in Africa, unlike other measures of emissions. We also include other major emitters.

We produce one final graph (figure 1.7), which measures per capita emissions as a function of per capita GDP. Per capita GDP is on the x-axis and per capita annual carbon dioxide emissions from energy are on the y-axis. The line represents the line of best fit: countries above the line use more fossil fuel energy to produce GDP than countries below it. There is a clear relationship between emissions (or energy use) and income. Rich countries emit more than poor countries. There are also some countries that are off the line of best fit: Norway has a high income and relatively low emissions, while Qatar has a middle level of income but very high emissions. The United States is just about on the line of best fit, while the European Union is somewhat below it.

Conclusion

We have covered a great deal of material in a short space, and it will be useful to summarize our main conclusions. The harm from climate change is expected to be much worse in poor regions, such as Sub-Saharan Africa, India and Southeast Asia, and small islands. The benefits from stabilization correspondingly help these countries the most, at least under most (but not all) scenarios. Both the harms and benefits are mostly in the future. It follows that future poor people, not present poor people, would gain the most from emissions reductions.

Rich countries have high per capita emissions; they also have contributed a great deal to the current stock of emissions, and they continue to contribute a great deal to the annual flow. But by the best measures, developing countries already contribute approximately the same to the flow and have not contributed much less to the stock, although they have a much lower per capita contribution to both the flow and the stock. Moreover, growth rates in emissions are much higher in developing countries than in developed countries. China is now the world's largest emitter in terms of flows of carbon dioxide. If growth patterns continue, China will overtake the United States as the largest contributor to the stock of carbon in the atmosphere.

Reducing emissions will be expensive. There are many low-cost reductions available using current technology, but long-term goals, such as zero emissions from the transportation and power sectors, will require large-scale technological improvements.

Given the scope of the harms and the cost of abatement, there is a clear case for significantly reducing emissions, because the total benefits of such reductions far outweigh the costs. To be sure, the precise nature of the reductions—their magnitude and timing—is disputed (and we will explore the reasons for the dispute). Nonetheless, there should be room for an agreement.

Why, then, is such an agreement not yet in place? A large part of the answer lies in the perceived costs and benefits for particular nations, including the United States and China. Another part of the answer involves competing claims about justice. Before turning to these points, it will be useful to see what methods have been considered for reducing emissions, and what exactly has happened to date.

Chapter 2

Policy Instruments

Many policy instruments are available for reducing greenhouse gas emissions. They include emissions taxes, tradable emissions permits (cap and trade), subsidies for emissions reductions, command and control regulations, labeling and information requirements, and subsidies for research and development of low-carbon technologies. In addition, nations will likely adapt to climate change by investing in measures that reduce the harmful effects of climate change. Adaptation measures include building seawalls that limit flooding caused by rising sea levels, moving from threatened areas, developing new plant varieties that survive extreme weather conditions, water conservation, improving responses to diseases that will spread as the climate warms, improving weather forecasting technology, and similar measures. There are as many combinations of mitigation strategies and adaption responses as there are analysts, with cap-and-trade systems and taxes dominating the current debate.[1]

Most of the work on the choice of emissions policies focuses on their relative efficiency—which policies can achieve reductions at the lowest possible cost. Because of the size of the problem presented by climate change, even modest improvements in efficiency can lead to large savings. Market-based mechanisms, such as taxes or cap-and-trade systems, hold the promise of reducing emissions at a much lower cost than other mechanisms. There are also, however, a number of underlying ethical issues that are important to the debate.

In particular, cap-and-trade systems might have different distributive effects than will taxes, particularly in the international setting.[2] The reason is that cap-and-trade systems might have the flexibility to achieve the desired distributive effects through the allocation of per-mits that taxes do not have.[3] If cap-and-trade regimes and taxes achieve similar environmental outcomes but cap-and-trade regimes have better distributional effects, there is a good case for cap and trade over taxes.

Another possibility is that taxes are unethical because they allow people to pay to pollute. They are like the sale of indulgences by the Medieval Church. People, particularly wealthy people, should not be able to cause harm to the rest of the world as they wish simply by choosing to pay money. Sometimes this argument is put in terms of environmental certainty. Cap-and-trade regimes provide a hard cap on total emissions and taxes do not. While people can always purchase permits in a cap-and-trade system, so in a sense they pay to pollute, this does not increase the total amount of pollution; no matter how much people want to pay, they cannot pollute more than the overall cap allows. Because they provide environmental certainty and because they do not allow people to pay to cause harm to others, cap-and-trade regimes, it is argued, are superior to taxes.

Our goal in this chapter is to examine these ethical arguments about the choice of climate policies. We will argue that neither ethi-cal argument has merit: cap-and-trade systems are not distribution-ally more flexible than taxes are, and they are not better at providing environmental certainty. We begin with a primer on the econom-ics of climate change and pollution control instruments and then turn to the ethical issues. We close with a brief discussion of optimal emissions reductions.

The Basic Economics of Climate Change

Climate Change as a Tragedy of the Commons

Climate change is perhaps the most important case of a widespread problem known as the tragedy of the commons: when people are given free access to a limited resource, they will tend to overuse it.[4] The reason is that their use, by depleting the resource, imposes costs on others, a cost that they do not consider because access is free. Worse, because others will use the resource if they don't, there is an incentive to race to the use of the resource first, before others do so. Open-access fisheries, common grazing land, and free road use are standard examples of the problem. In each of these cases, use by one person imposes a cost on others by depleting the resource. If access is free—users do not have to pay for this cost imposed on others—they will tend to overuse the resource.

Climate scientists have taught us that the atmosphere is a limited resource similar to roads or fisheries—the atmosphere can only safely absorb a limited amount of carbon dioxide. Therefore, whenever people engage in activities that emit carbon, such as heating, cooling, transportation, or the use of metals, paper, cement, chemicals, or meat, they deplete the resource but do not pay a price for the harm they impose on others. The result is overuse. In economic terms, carbon emissions are an "externality"—the effects of emissions on other people are not included in the price. The price people pay—zero—is below the true cost, so people use too much.

There are two broad classes of approaches to the problem. The first, and traditional, approach for environmental problems is to simply order polluters to pollute less, an approach known as command and control regulation. This can be done by limiting the total amount of an activity or by mandating the use of particular technologies. In the climate context, for example, we might reduce carbon emissions by mandating that automobile fleets have certain average gas mileage, by mandating that biofuels be mixed with gasoline, by mandating that buildings have a minimum amount of insulation, or by mandating that power be produced with renewable fuels. There are a large number of technologies that might be used to limit emissions, and the government could decide on the mix of these technologies.

The second approach is to rely on individuals and markets to choose the mix of technologies to reduce emissions. Cap-and-trade regimes are a classic example. In a cap-and-trade regime, the government sets an overall limit on emissions but lets people choose the method of meeting that limit. In particular, the government issues pollution permits representing the total amount of pollution that will be tolerated and allows people to buy and sell the permits. Anyone who can reduce emissions at a cost less than the market price of the permits will do so. For example, suppose that permits trade at $50 per ton of carbon dioxide. If you are currently emitting carbon dioxide, you have to buy permits. If you can reduce emissions for less than $50 per ton, you would do that instead of buying permits. Only emissions that cannot be reduced at the market price of the permits would remain. Because everyone would face a common price for permits, people would seek out and use the emissions reduction opportunities that cost less than the permit price. Individuals and the market would choose the reduction technologies and methods. All the government needs to do is choose the overall amount. By harnessing information held by individuals and by the market to choose the mechanisms for reducing pollution, these systems can significantly reduce the costs of emissions reductions.

This approach was most famously used by the Clean Air Act in regulating sulfur dioxide emissions, and studies have shown that it reduced the costs substantially.[5] For this reason, market-based mechanisms such as cap-and-trade regimes are generally considered more promising than command and control systems. This is particularly true in the climate context. Because the sources of emissions are so pervasive and because of the wide variety of emissions reduction technologies, the chance that the government could correctly choose the right sources for reduction and the right technologies is low. Reliance on market-based mechanisms in the climate context is likely to be especially important.

We can illustrate the effects of a cap-and-trade system using a simple supply and demand diagram. The x-axis of a supply/demand graph is the quantity of a good. Here the good is abatement, the reduction in emissions. Thus, as we move to the right on the x-axis, we have fewer emissions. The y-axis is dollars spend on abatement.

We need to draw curves representing the marginal benefits and costs of reducing carbon emissions (i.e., the incremental benefits and costs from reducing emissions by one more unit). The marginal benefit curve slopes downward. The reason is that the benefit of reducing emissions ever further will go down as emissions go down: keeping total concentrations of carbon below some very high number is likely very important, but reducing concentrations from a low

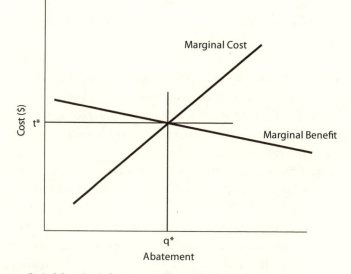

Figure 2.1 Marginal Cost and Benefit of Abatement.

number to a very low number is probably not all that important. The benefit of reducing emissions by one more unit, which is what the marginal benefit curve represents, goes down as carbon concentrations go down.

Calculating the marginal benefit of reducing emissions is enormously difficult. There is a large amount of uncertainty about the impacts of emissions. Moreover, even given best guesses about the impacts, the calculation involves numerous ethical issues, such as how we compare harms of various sorts. We discuss a number of these in later chapters, such as how to count harms that occur in different periods of time. We assume here that such a calculation can be done.

Similarly, the marginal cost of abating is upward sloping, reflecting the fact that abating a little is less costly per unit than abating a lot. Figure 2.1 illustrates this result.

The government will want to set the cap on emissions where the marginal cost of reducing emissions equals the marginal benefit. If the cap is set so that marginal cost is lower than the marginal benefit, reducing emissions by another unit would cost less than the benefit, so we should have a tighter cap. If the marginal cost is higher than the marginal benefit, we could loosen the restriction and save more money than we lose in benefits from less pollution. In the graph, the government will want to set the cap at q^*.

If there is trading in permits, they would trade at a price equal to t^*. To see why, suppose that t^* (the marginal cost of abating by q^*) is equal to $50. If permits were to trade at a higher price, say $60, people could reduce emissions rather than purchasing the permits. If permits were to trade at less than $50, people would purchase the permits rather than reduce emissions.

Finally, if the government were to auction the permits rather than give them away, the auction would raise an amount of money equal to the rectangle with the corners at q^* and t^*. Just as the permits would trade in the market at t^*, they would sell for t^* in the initial auction. Because the government would issue q^*, the total amount raised would be the product, q^*t^*. For example, if the trading price were $50 per ton of carbon dioxide and we issued permits for 6 billion tons, the auction would bring in $300 billion.

Taxation is an alternative to a cap-and-trade system. Suppose that instead of issuing permits, the government imposed a tax on all carbon emissions equal to t^* ($50 in our example). Anyone who could reduce emissions at less than this amount would do so, and anyone who could not reduce emission at that cost would have to pay the tax. The total amount emitted would be where t^* intersects the marginal cost curve, which is q^*. That is, the total amount emitted is the same under a tax set at t^* and a cap-and-trade regime with a limit of q^*. In both cases, individuals and the market would choose the technologies and methods of reducing emissions. Moreover, the tax raises revenue equal to t^*q^*, which is the same amount raised by a cap-and-trade regime if the permits are auctioned rather than given away.[6] In their very basic form, the two are equivalent.

Subsidies are a third option for creating incentives to reduce emissions, an option that is already widely used. For example, the United States offers a large number of subsidies for clean energy, such as a tax credit for wind energy.[7] In theory, subsidies can create incentives that share some of the features of taxes and permit systems. In a tax system, polluters save money by reducing emissions and paying less tax; in a cap and trade, by purchasing fewer permits; and in a subsidy system, by getting a subsidy. For example, if the tax is $50 per ton, polluters will eliminate emission to the extent that they can do so for less than $50; they will have to pay tax on the rest. If the subsidy is $50 per ton, polluters would reduce emissions and receive the subsidy only if they can reduce emissions for less than $50 per ton. If it costs $60 per ton to reduce emissions, they would not do so in order to receive a $50 subsidy.

Subsidies, however, do not offer all the benefits that taxes or permits do. To illustrate, consider a subsidy for renewable energy. Although such a subsidy would make renewable energy less expensive as compared with fossil fuel, it would not increase the price of fossil fuel. Therefore, people might switch from fossil fuel to renewable energy but would not have an incentive to reduce overall energy use, unlike with a tax or a permit system.

Moreover, subsidies present administrative difficulties that taxes and permits do not. To calculate a subsidy, the government must be able to determine the amount that would have been emitted absent the subsidy. That is, the subsidy is based on the difference between the amount that would have been emitted and the amount that is actually emitted. There is no straightforward way to calculate this difference—once we have a subsidy in place and a polluter reduces emissions, we can no longer see what they would have done without the subsidy. In the climate change context, people refer to this as the problem of additionality—the government does not know whether the emissions reductions due to the subsidy are in addition to the reductions that might have taken place anyway. Taxes and cap-and-trade regimes do not have this problem because they are based on actual emissions, not emissions reductions from a theoretical baseline.

The subsidies put in place as part of the Kyoto Protocol illustrate the problem of additionality. The Kyoto Protocol imposes emissions limits on developed countries but, because of concerns about imposing costs on poor countries, it imposed no limits on developing countries such as China and India. Instead, to achieve reductions in these countries, the Kyoto Protocol uses a subsidy known as the Clean Development Mechanism, or CDM. Under the CDM, polluters in developed countries can claim emissions reductions credits for projects in the developing world that reduce emissions below what they would have been. It is the single largest market-based atmospheric pollution regime ever put in place.[8] The CDM, however, has had terrible problems because of additionality.

For example, production of a common refrigerant, known as HCFC-22, produces a by-product, HFC-23, which is a very potent greenhouse gas (with a global warming potential 11,700 times that of carbon dioxide). Because of this very high global warming potential, the CDM subsidy for capturing and destroying HFC-23 far exceeds the value of the HCFC-22 produced. There is, therefore, an incentive to create refrigerant plants solely to produce and then destroy the HFC-23 and receive CDM credits. To illustrate, imagine that it costs 100 to build a plant that produces HCFC-22 that is, to take an extreme case, worthless. As a by-product of producing HCFC-22, it produces HFC-23, which has a high global warming potential. Because HFC-23 is so potent, a company seeking CDM credits, might pay the builder more than 100 to shut the plant down, creating an incentive to build the plant merely to have it shut down. In effect, building and shutting down these plants is simply a method of generating CDM credits. Because we cannot tell whether a plant would have been built absent the subsidy, we cannot tell whether any given destruction of HFC-23 actually reduces emissions or whether it is just a game to capture the subsidy. The CDM is an example of distributive judgments producing bad climate policy.

In more complex settings, taxes, permits, and hybrids behave differently. There is a large literature on choice of instruments in various more realistic settings. Much of the discussion is technical economics and is not central to the ethical issues we focus on here. There are, however, two ethical claims about instrument choice: that cap-and-trade systems are likely to have better distributive consequences and are also likely to have better environmental consequences.

Distributional Differences

Some claim that cap-and-trade systems are more flexible than taxes, allowing us to better tailor the distributive consequences. The reason is that governments can choose how to allocate the initial permits.

If, for example, the government wants to help poor people, it can give them permits, which the poor people can then sell to polluters. The initial allocation does not affect who will ultimately own the permits, as they will be traded in the market and those who value the permits most will end up owning them regardless. This means that the initial allocation does not affect the efficiency of the program because polluters will face an incentive to reduce emissions at less than the trading price of the permit regardless of who initially holds them. Therefore, the initial allocation of permits in a cap-and-trade regime offers a unique opportunity to improve the distribution of resources without efficiency losses (aside from the cost of redistributing). (Normally, redistributing more by, say, increasing marginal tax rates at the top produces inefficiency because people who are taxed at high rates will work and save less.) We refer to this in later chapters as *ex post* efficiency. This effect is even stronger in the international context. If we allocate permits to poor countries, polluters in wealthy countries will have to purchase them from poor countries, transferring resources from wealthy countries to poor countries while maintaining the efficiency of the system. The Stern Review, reflecting this logic, concluded, "[a] major advantage of emissions trading schemes is that they enable efficiency and equity to be considered separately."[9]

The potential redistribution is quite large. The United States currently emits about 7 billion metric tons of carbon dioxide per year. If permits traded at $50 per ton under a cap-and-trade system that covered about 60 percent of emissions, the total cost of enough permits to cover U.S. emissions would be over $200 billion per year. If most of these had to be purchased from foreign countries, the United States would, in effect, be transferring most of that money to foreign countries. The net outflow could easily be much higher if permits were more expensive or more emissions were covered. To get a sense of the size of this number, a rough estimate of current U.S. foreign aid is about $20 billion per year.[10]

Within any single country, the argument that permits have more flexibility than taxes is not correct. If permits are auctioned and the tax is set at the price that permits would sell for, the two are basically identical. Polluters would pay the same amount to the government, and the government could spend the money the same way under either system. If some or all permits are given away, we can adjust the tax system to provide exemptions or credits to the same individuals or industries that would have received free credits. For example, if we gave away the permits on a per capita basis, we could achieve a similar effect by refunding all tax revenues on a per capita basis. Any pattern of permit allocations can be matched with an allocation of tax revenues. Moreover, the efficiency effects would be the same: regardless of how tax revenues are spent, polluters would face an incentive to reduce emissions if they can do so at less than the tax; just like with permits, regardless of how they are initially allocated, polluters have an incentive to reduce emissions if they can do so at less than the trading price.

In the international context, the issue is more subtle. Normally, each country imposes taxes on activities within its borders and keeps the revenue (aside from relatively small transfers, such as foreign aid). Countries with large emissions would suffer the consequences of a tax—they would end up incurring costs to reduce emissions—and would also keep the tax revenue. In a permit regime that applied across many countries, permits would have to be allocated to each country as part of the treaty establishing the permit regime.[11] The individual countries would then decide what to do with the permits, either giving them to whomever they wish such as favored industries, auctioning them off domestically or internationally (and using the money for whatever purpose the government desired), or some combination. The allocation of the permits would determine the distributive effect. Because permits can be allocated in many ways, they can achieve a wide variety of distributive goals. For example, permits can be allocated to mimic the flow of tax revenue under a carbon tax. This would involve allocating them in proportion to existing emissions. They can be allocated on a per capita basis, which we discuss in chapter 6. They could be allocated based on any criteria imaginable.

In the single country context, we noted that the government can simply allocate tax revenues instead of initial permits to achieve the desired distributive effects. This is also true in the international context, but now it would involve transferring large sums of cash—tax receipts—to poor countries. This, it is claimed, is unlikely. We would not agree to increase our foreign aid budget by tenfold merely because we are entering into a climate treaty. Even though we would not allocate tax revenues this way, we might, it is claimed, allocate permits to exactly the same effect.

The difference is purely one of framing—it is unusual to see countries sharing tax revenues, but because permits would be new, various initial allocations do not seem odd.[12] It seems natural that each country would keep its own tax revenues and it seems natural to many that permits should be allocated on a per capita basis (on the theory that each person has an equal right to the atmosphere) or in a way that helps the poor. Given this framing, the claim is that there will be more flexibility in permit allocations than for taxes. By framing the transfers as permits rather than cash, we would be willing to transfer more.

While we cannot dismiss this argument outright, it seems implausible. It is likely that in some cases, the framing of regulatory policies masks their distributive effects, but the scale of the transfers in the climate context would make it almost impossible to hide the effects. That is, the flexibility claim is that rich countries would be more willing to give permits to poor countries only to purchase them back for cash than they would be to transfer the cash directly. When we are talking about hundreds of billions of dollars a year, amounts that greatly exceed existing foreign aid budgets, there is no chance that people will be fooled. Either they will want to give this money to poor countries as part of a climate treaty or they will not. Framing the issue as permit allocations will not fool rich countries into agreeing to vast increases in their transfers to poor countries.

Moreover, transferring money to poor countries is complicated by concerns over how the money will be spent. Foreign aid and grants by international institutions have substantial restrictions. If rich countries were to give large amounts of aid through permits to poor countries, it seems likely that they would want similar restrictions. It would not be easy, however, to direct permits the way that foreign aid is now directed, with very particular spending and oversight requirements. Instead, permits would likely have to be given to governments around the world which would use them as they see fit. Thus, for permits to lead to more transfers to poor countries, rich countries would have to agree to increase their transfers by a large multiple while at the same time reducing the control they have over the use of the money. This seems unlikely.

Purchasing the Right to Pollute

The second ethical claim about the choice of pollution control instruments is that taxes are unethical because they allow people to pay to pollute. They send a message that pollution is fine so long as one pays the fee, like going to see a movie or buying a soda. But many argue that emitting carbon damages the planet, and it is not fine to simply pay to do it. If you are in a nonsmoking car of a train, it is not okay to light up a cigarette on the theory that you can afford the fine. As mentioned, some have analogized carbon taxes to the sale of indulgences by the Medieval Church. Perhaps the best version of the argument, made in the closely related context of the purchase of carbon offsets,[13] is a tongue-in-cheek website known as Cheat Neutral (found at www.cheatneutral.com). If you cheat on your spouse, you can buy cheat offsets from the website, which promises to use the funds to pay other couples not to cheat. By purchasing offsets, you ensure that the total amount of cheating is held constant; indeed, by purchasing additional offsets, you could even reduce total cheating notwithstanding your extracurricular activities.

A related claim is that because taxes allow people to pollute as much as they want by paying the tax, taxes do not provide the necessary environmental certainty—they run the risk of carbon concentrations that are dangerously high. Only a fixed cap on emissions ensures that we keep concentrations at a safe level. As two well-known climate analysts put it, "a cap-and-trade system, coupled with adequate enforcement, assures that environmental goals actually would be achieved by a certain date. Given the potential for escalating damages and the urgent need to meet specific emission targets, such certainty is a major advantage."[14] Or as another analyst put it, "one of the great advantages of tradeable permit schemes (or at least cap-and-trade tradeable permit schemes) is their environmental certainty. Relative to all other environmental policy instruments they provide—assuming perfect monitoring and complete enforcement—complete certainty with respect to the total level of emissions."[15]

The environmental certainty claim has a number of problems. Suppose that there was a hard cap on global emissions with strong enforcement measures to ensure compliance so that we knew that carbon concentrations would be limited to a chosen amount. This would create emissions certainty but would not create environmental certainty. The reason is that we have very little understanding of the environmental outcomes for any given level of carbon concentration. As we noted in chapter 1, the IPCC, for example, puts climate sensitivity (the equilibrium global average temperature increase for a doubling of carbon dioxide concentrations) between 2°C and 4.5°C, a range wide enough to include modest but easily manageable harms to severe disruption. That is, even if we knew for certain that carbon dioxide concentrations would at most double, we would have very little idea of the environmental outcome.

Moreover, modest changes in carbon dioxide concentrations do not substantially change our expectations for the environment. The International Energy Agency compared a hard emissions cap to policies that allowed some flexibility.[16] In particular, it compared a hard cap that cut emissions in half by 2050 to a cap with the same goal but that put a ceiling and floor on permit prices, so that if, say, permits traded above some amount, polluters could purchase additional permits at that price, effectively converting the cap into a tax. The hard cap fixed concentrations at 462 ppm while the flexible policy produced a range of outcomes between 432 ppm and 506 ppm. The environmental outcomes in the two cases were essentially identical: the median temperature increase was 2.49 for the hard cap and 2.53 for the flexible policy; the risk of avoiding a very bad outcome (5°C) was 98.5 percent for the hard cap and 98.3 percent for the flexible policy. But the flexible policy cost less than one-third of the hard cap. It concluded, "achieving a given concentration level (such as 462 ppm) exactly or on average does not make any real difference to the environmental outcome. The uncertainty introduced by price caps in concentration levels is entirely masked behind the uncertainty on climate sensitivity."

The best argument for environmental certainty is that there might be a so-called tipping point, a level of concentration at which damage gets dramatically worse. Various environmental outcomes might be nonlinear, so that we see very little change until carbon concentrations hit some level, at which point we see dramatic and fast changes. For example, as sea-ice melts, it exposes a darker ocean surface which absorbs more heat, amplifying the warming. If this effect is strong enough, sea-ice melting might be self-sustaining once it gets past a given point. Scientists looking at climate history going back millions of years see evidence for very fast changes, creating real concern about the possibility of a tipping point.[17] If there is a tipping point, it would be very important to keep concentrations below that level.

The problem with this argument is that we have no idea what level of concentration (if any) creates a tipping point. This means that we cannot set a hard cap knowing that doing so avoids the problem. Concerns about tipping points argue for an overall higher level of environmental stringency as a means of buying insurance against very bad outcomes, but they do not argue for any particular level of concentration. William Pizer modeled the effect of tipping points on the choice between taxes and permits and concluded that if we were near a tipping point, the differences between the two instruments is swamped by the sheer necessity of putting in place a stringent regime quickly.[18] That is, if we were near a tipping point, it wouldn't matter so much how we reduced emission as it would matter that we did so quickly. Moreover, there may be substantial harms from setting policy based on incorrect guesses about a tipping point.

The second problem with the environmental certainty claim is that it makes unrealistic assumptions about how taxes and permits would work. It assumes compliance with a cap that remains fixed over time regardless of cost, and it assumes no adjustment to the tax rate if emissions exceed expectations. Once we relax the assumption of compliance with a cap that does not change, we lose any benefit of certainty that a cap might offer. For example, if costs under a stringent cap turn out to be very high and we therefore loosen the cap, we no longer have certainty over final concentrations. Similarly, once we relax the assumption of a tax that is not adjusted to take its effect on behavior into account, taxes are able to achieve more certain carbon concentrations. Thus, if a given level of tax does not produce the predicted emissions reductions, we can increase the tax rate. Realistically, both caps and taxes will (and should) be adjusted over time as we learn more about climate science and the costs of reducing emissions.

The claim that paying a carbon tax to be allowed to pollute is like paying an indulgence to sin is even more tenuous. No serious policy would reduce emissions to zero, at least not until the distant future. The question is how best to reduce emissions. If a well-designed tax can reduce emissions to an appropriate level more cheaply than other policies, it is hard to characterize it as allowing people to pay to behave worse than they would under other policies. For example, if a tax is designed to limit concentrations to a specified level, people would pay a tax when they emit carbon and reduce concentrations consistent with the desired goal. If a cap-and-trade regime were designed to achieve the same target, people would similarly pay to purchase permits to pollute and would reduce concentrations to the specified level. Once we have agreed to a market-based system rather than an absolute ban, we have agreed that people can pay to pollute.

The argument breaks down because emitting carbon is part of everyday life. Unlike nonsmoking policies or Church policy toward sin, the goal of climate change policy is not simply to stop emissions cold. A carbon tax forces polluters to bear the costs of emissions, and that is enough.

A Note on Optimality and Cost-Benefit Analysis

Throughout this book, we will refer to an "optimal" climate treaty. We do not have any particular treaty obligations in mind when using this phrase. Instead, we mean primarily that the treaty reduces emissions to the extent that the cost of reducing emissions a little bit more is equal to the benefit of reducing emissions a little bit more; in economic terms, the marginal costs equal the marginal benefits.

If this is not true, we can reduce emissions a little bit more and get a benefit that is greater than the cost or reduce emissions less and save more money than the costs of additional harms (fewer benefits) to the climate.

We also mean that in determining the marginal costs of reducing emissions and in determining the harms from emissions, we consider the world as a whole—we want to find the lowest cost emissions reductions wherever they may be and to count all of the harms from climate change. This is important because if we omit low-cost sources of emissions reductions, the costs will be higher than otherwise, and we would miss opportunities for emissions reductions where the marginal cost is less than the marginal benefit.

One important point to note is that in computing harms, analysts sometimes add distributional weightings so that the harm to a poor person is counted more than it otherwise would merely because the person is poor. The reason is that helping a poor person in a sense has a double benefit: it helps the person and it reduces inequality. When we use the notion of optimal emissions, we do not include these weightings because we consider distributional issues separately.

The procedures for computing marginal costs and benefits are complex and will be particularly so in the climate change context because of the scale of the problem and the uncertainty. There are heated ethical and methodological debates about the merits of various procedures, such as trading off lives for money or the environment for money. Valuing future contingencies, determining the existence value of species, and adequately accounting for risk are all difficult. We cannot resolve these issues here, but we mention two that might be particularly important in the climate context.

The first issue is that some argue environmental degradation cannot be given an accurate value, particularly degradation of the scope involved in the worst cases of climate change. Environmental harms might be said to be incommensurate with other types of harms. If so, we cannot simply plug all of the numbers into a giant optimizing machine, weighing one type of harm against another.

In evaluating this argument, we need to distinguish between a claim that cost-benefit analysis often undervalues the environment from a claim that the environment simply cannot be valued at all (or that it has infinite value). There is a substantial field of research that attempts to put a value on environmental services—the often unseen benefit the environment provides.[19] This research is not yet widely incorporated into climate change cost-benefit analysis, and it needs to be. If the results of this research show that the environment is often undervalued, cost-benefit analysis needs to be adjusted to use better valuations. In the case of severe climate change, environmental services that we take for granted now, and are a small part of the economy, may become scarce and very valuable. Agriculture is a small percentage of the U.S. economy now, and few people need to work on farms. If climate change is severe, however, feeding the population might be much more difficult, and the marginal cost of agricultural degradation might go up. Proper valuation of the environment in these cases may significantly change the conclusions from cost-benefit analysis of climate change. When we use the notion of an optimal climate treaty, we include full valuation of the environment.

Yet many argue that environmental harms have special moral status. If it becomes significantly hotter in India, and there is considerable hardship as a result, can cash payments really make Indians better off than they were before the temperature increase? Some welfarists think that deprivation of some goods, such as literacy, cannot be adequately compensated with cash. Deontological theorists are even more inclined to be skeptical of the claim that all human goods can easily be traded off against one another. For example, arguments for various rights typically view rights as incommensurable; under such theories, it is not ethical to take a right by force and then offer cash as compensation. In the environmental context, intuitions on behalf of these arguments are sometimes based on bad experiences with compensation, such as with the medieval sale of indulgences.[20] Environmental harms are particularly subject to these sorts of claims because of the special consideration many give to the environment.

It is not clear what to make of these claims in the climate change context. At some point, nations will have to decide how much to reduce emissions. If the incommensurability claim is taken to mean that the harms to environment have infinite value, the only course of action would be to stop emitting all greenhouse gases immediately. Needless to say, this view does not command much assent. Short of this, however, a decision of how much and how fast to reduce emissions includes an implicit valuation of environmental harms. The argument is about the proper value rather than a claim that the environment cannot be valued at all. Even if goods are in some sense incommensurable—even if there is no simple or easy metric along which to align them[21]—it remains true that trade-offs must be made among these very goods.

A second concern with the use of cost-benefit analysis in the climate context is that many of the harms may be irreversible. A lost species or habitat can never be replaced. Cost-benefit analysis that does not consider irreversible harms may produce incorrect prescriptions. The issue is more subtle than it first appears, however. Considerations of irreversibility might not lead to the acceleration or strengthening of climate actions to prevent such permanent harms. The reason is that there are irreversible harms on both sides. If a country opts to cut back on emissions at the cost of economic growth, then some people will die or suffer in ways that they otherwise would not have. These harms are irreversible just as lost species and habitats are irreversible. Exactly which way considerations of irreversibilities cut is unclear, but regardless, when we say we want to reduce emissions by the optimal amount, this is to be calculated using the best possible procedures including full consideration of irreversibilities.

• • •

The best policies are those that reduce emissions at least cost, taking into account administrative considerations such as transparency. None of the policy options under serious consideration have significant ethical advantages over the others.

Chapter 3

Symbols, Not Substance

Climate change has attracted an enormous amount of attention around the world and has generated immense activity in the form of legislation, regulation, diplomatic negotiations, international conferences, and meetings, but to date, this activity has yielded very little of substance. Most of the legislation and other forms of government action that have taken place have been largely symbolic. This is particularly true of the government measures taken so far in the United States, most of them at the state and local levels. More significant measures, notably the European Union's Emission Trading System, have produced far less than advertised, and are wholly inadequate by themselves to address the climate problem. Symbols, not substance, have been the order of the day.

And this speaks just of the developed world. In the developing world, arguments that poor countries should cut back on greenhouse gas emissions have been met with hostility. These countries have done little, even symbolically, to cut greenhouse gas emissions, and indeed over the last twenty years, the growth of global emissions has occurred mainly because of the economies of the developing world. Rapidly developing countries fear the political instability that could occur if emission limits suddenly derailed economic growth. Some of these countries, such as China, India, and Brazil, finally can imagine western living standards for many of their citizens and the kind of global influence that has long been denied to them despite their size and importance—and now western countries tell them that they are growing too fast. Many other developing countries are simply very poor and fear the consequences of higher energy prices for their people. In recent months, China has persuaded some commentators that it looks forward to a world with lower greenhouse gas emissions and plans to profit by investing in green technology. But so far China's statements are just words, and its government has not yet shown a serious commitment to reducing greenhouse gas emissions.

Politicians resort to symbolism to meet a short-term demand for action that they believe will produce long-term costs for which they will be held politically responsible. Their behavior up to the present suggests that they believe the public does not understand the economic cost of meaningful climate limits and will punish them if economies sour as a result of government action to counter climate change. The rift between north and south exacerbates this problem. The public in developing countries has been encouraged to believe that climate change mitigation is the responsibility of the north, and developed countries should pay for the remedy. Thus, in the south the belief exists that countering climate change should be costless—that is, costless for people in developing countries because the costs will be paid by the north. This view is simply not realistic, as we will discuss in subsequent chapters.

Symbolic behavior is not always futile. One symbolic action can lead to another, and as more and more people take actions with similar goals, a sense may develop that enough people care about a problem to make substantive government action possible. Support for symbolically potent but environmentally trivial municipal laws requiring developers to plant gardens on roofs, then, may help spread the word about climate change and convince people around the world that enough people care about this problem to make significant emission limits possible. The problem is that people sometimes believe symbolic behavior in itself is enough and are thereby lulled into a sense of complacency that the issues surrounding climate change have been addressed. Or symbolic behavior might just be the best that politicians can do when they realize that the public will not tolerate the costs that would be incurred to address climate change adequately. The record so far is not encouraging.

The Framework Convention

Although the science of climate change has been developing for some time, international action on climate change began only a few decades ago. After some preliminary actions in the late 1980s, more than 180 nations, including the United States, signed the Framework Convention on Climate Change at the 1992 Earth Summit in Rio de Janeiro. Its stated objective was the "stabilization of greenhouse gas concentrations in the atmosphere at a level that would prevent dangerous anthropogenic interference with the climate system." This goal is to be achieved in a time frame "sufficient to allow ecosystems to adapt naturally to climate change, to ensure that food production is not threatened, and to enable economic development to proceed in a sustainable manner."

To achieve these goals, the Framework Convention incorporates several different principles of justice. Under the principle of common but differentiated responsibilities, developed nations are to be responsible for a larger share of emissions reductions (or possibly all reductions) because of their greater wealth. The Framework Convention also includes backward-looking language, noting that "the largest share of historical and current global emissions of greenhouse gases has originated in developed countries . . . per capita emissions in developing countries are still relatively low and . . . the share of emissions originating in developing countries will grow to meet their social and development needs."

Notwithstanding these goals, the Framework Convention is almost entirely symbolic. It contains no binding limits on emissions. Instead, it contains only aspirational language such as that quoted above calling for stabilization of emissions to prevent dangerous interference with the global climate. Industrialized nations did agree to try to reduce emissions to 1990 levels by 2000, but this agreement was not made a legally binding obligation. Developing nations made no commitments. The aspirations appear to have had no effect. Since 1992, emissions have increased globally by more than 30 percent. Annex I nations (developed nations), as a group, have increased their emissions over this period by 6 percent rather than reduced them. Meanwhile, emissions from non–Annex I nations (developing nations) have skyrocketed by more than 75 percent during this period.

It would be a mistake, however, to say that the Framework Convention is not important. It is a significant achievement. Perhaps the most important element of the Convention is that it requires developed nations to produce annual inventories of their emissions. These inventories have helped us understand the causes of climate change and are likely to form the basis of future abatement obligations. Moreover, the procedures developed in the Framework Convention began the process that eventually led to the Kyoto Protocol and remain the basis of negotiations today.

The Kyoto Protocol

The Kyoto Protocol was adopted by the parties to the Framework Convention in 1997 and went into force in 2005. The Protocol imposed binding quantitative limits on greenhouse gas emissions for industrialized nations listed in Annex I. The forty Annex I nations were expected to collectively reduce emissions by 5 percent from 1990 levels. However, different nations faced different limits. The United States, if it had ratified the treaty, would have had to reduce emissions by 7 percent. The European Union was obligated to reduce emissions by 8 percent; under a separate article of the Protocol, European governments were permitted to allocate different shares of this reduction obligation to different EU members. Russia and New Zealand had to maintain emissions at their 1990 levels—that is, reduce them by 0 percent. Australia was required to ensure that its emissions did not increase by more than 8 percent. Parties were required to meet their limits sometime between 2008 and 2012, depending on the country. The Protocol also put in place an international cap-and-trade system.[1]

The Kyoto Protocol did not impose quantitative limits on greenhouse gas emissions for developing countries. Their main roles in the treaty regime were as beneficiaries of the Clean Development Mechanism. Under this system, Annex I nations could increase their greenhouse gas allotments by funding projects in developing nations that reduced greenhouse gas emissions. As we noted in chapter 2, the CDM has attracted some serious criticism. Because it is difficult to know whether an energy-efficient power plant (for example) would have been built in the absence of financial assistance from the developed world, the CDM may actually increase global emissions by giving developed nations a relatively cheap escape hatch from their limits.

Today, nearly every country in the world belongs to the Kyoto Protocol; the United States stands out as the only major country that has refused to ratify that treaty. The other major holdout, Australia, ratified the Protocol in 2007. For this reason, the United States has suffered a great deal of opprobrium. Many people believe that in the Kyoto Protocol nations agreed to sacrifice their interests for the sake of global well-being, and American intransigence represented national selfishness at its worst. The truth is more complex. The Kyoto Protocol did not, in fact, require sacrifices of any significance on the part of any country—except for the United States—and even then, its parties have had trouble living up to their limited burdens.

As already noted, the Kyoto Protocol imposed no quantitative limits on the vast majority of countries in the world—all the countries (mainly developing countries) not included in Annex I. These unburdened countries include China, the world's largest emitter of greenhouse gases, and other significant emitters of greenhouse gases, such as South Korea. Even among the Annex I countries, the limits are less than meets the eye. Russia agreed to keep emissions to 1990 levels (a 0 percent reduction), but its economy had shrunk so much during the 1990s that this limit was essentially meaningless. Germany had experienced an economic contraction as a result of the absorption of the former East Germany, and Britain's greenhouse gas emissions were independently falling because of a shift from coal to natural gas in its energy sector.[2] Because the EU countries could allocate among themselves the responsibility for reaching the EU's 8 percent target, the other countries enjoyed the slack that resulted from the "windfalls" in Germany and Britain.

By contrast, the United States would have experienced quite substantial costs if it had ratified the Kyoto Protocol—between 50 percent and 80 percent of the global burden.[3] The negative reaction in the United States should be understood with this fact in mind. In 1997, the Senate unanimously issued a resolution opposing greenhouse gas emissions limits that harmed the U.S. economy or that were not shared by developing countries.[4] The Senate complained specifically (and accurately) that a climate change agreement that did not include developing nations is "environmentally flawed." Although the United States signed the Protocol in 1998, the Clinton administration did not submit the treaty to the Senate for approval. In the face of unanimous, bipartisan opposition, such a move would have been fruitless. The Bush administration's more explicit opposition to the Kyoto Protocol in 2001, then, merely confirmed the status quo as it stood in the United States at that time.

We have observed that the other Annex I nations were not subject to obligations under the Kyoto Protocol that were as burdensome as those placed on the United States. Nonetheless, their compliance with their limited obligations has been spotty. Consider first the members of the European Union.[5] Of the twenty-five members with targets, only about half are in compliance (table 3.1).[6] Others may well come into compliance but the overall record is not encouraging. Indeed, putting aside Germany and Britain, which experienced emissions reductions for reasons other than climate policy, emissions in the EU actually increased on a per capita basis from 1990 to 2004, leading two scholars writing in 2007 to conclude that "few, if any, countries have adopted effective climate policies to date."[7] This conclusion is reinforced by the record of the remaining Annex I countries, shown in table 3.2, which suggests that the best prediction of compliance is not the adoption of wise climate policies but (involuntary) economic collapse.[8]

The Bali Road Map

In 2007, the nations of the world met in Bali to develop an additional agreement. The positions and divisions of the various actors were remarkably similar to those in Kyoto. Perhaps most important,

Table 3.1
Compliance Records of EU Countries

Country	Target	1990–2006	Compliant?
EU15			
Austria	–13.0%	15.2%	No
Belgium	–7.5%	–6.0%	No
Denmark	–21.0%	1.7%	No
Finland	0.0%	13.1%	No
France	0.0%	–4.0%	Yes
Germany	–21.0%	–18.5%	No
Greece	25.0%	24.4%	Yes
Ireland	13.0%	25.5%	No
Italy	–6.5%	9.9%	No
Luxembourg	–28.0%	1.2%	No
Netherlands	–6.0%	–2.6%	No
Portugal	27.0%	38.3%	No
Spain	15.0%	49.5%	No
Sweden	4.0%	–8.9%	Yes
UK	–12.5%	–16.0%	Yes
Other EU			
Bulgaria	–8.0%	–46.2%	Yes
Cyprus		66.0%	
Czech Republic	–8.0%	45.0%	No
Estonia	–8.0%	–55.7%	Yes
Hungary	–6.0%	–31.9%	Yes
Latvia	–8.0%	–55.1%	Yes
Lithuania	–8.0%	–53.0%	Yes
Malta		45.0%	
Poland	–6.0%	–28.9%	Yes
Romania	–8.0%	–43.7%	Yes
Slovakia	–8.0%	–32.1%	Yes
Slovenia	–8.0%	1.2%	No

Table 3.2.
Kyoto Protocol Compliance as of 2006: Other Annex I Countries

Country	Target	1990 and 2005	Compliant?
Turkey	N/A	74.4%	N/A
Belarus	–8%	–40.6%	Yes
Liechtenstein	–8%	17.4%	No
Monaco	–8%	–3.1%	No
Switzerland	–8%	1.7%	No
United States	–7%	16.3%	N/A—refuses to ratify
Canada	–6%	25.3%	No
Japan	–6%	6.9%	No
Croatia	–5%	–5.4%	Yes
New Zealand	0	24.7%	No
Russian Federation	0	–28.7%	Yes
Ukraine	0	–54.7%	Yes
Norway	1%	8.8%	No
Australia	8%	25.6%	No
Iceland	10%	10.5%	Almost

the developing countries, above all China and India, resisted any commitment to emissions reductions. The United States took a broadly similar approach, insisting on developing country participation and also resisting firm commitments. European nations, by contrast, sought firm targets, above all from the United States.

What emerged, the Bali Action Plan or "road map," reflects an uneasy compromise among the contending forces—and represents a clear victory of symbol over substance. The preamble recognizes "that deep cuts in global emissions will be required to achieve the ultimate objective of the Convention." However, no such cuts are mandated. In terms of what has been "decided," a clear division separates developed from developing countries. Developed countries agree to "consideration of [m]easurable, reportable and verifiable nationally appropriate mitigation commitments, including quantified emission limitation and reduction objectives . . . taking into account differences in their national circumstances." By contrast, developing countries agree to "consideration of [n]ationally appropriate mitigation measures . . . in the context of sustainable development, supported and enabled by technology, financing and capacity-building in a measurable, reportable and verifiable manner."

For our purposes, three points are worth emphasizing. First, no nation has agreed to anything. Both developed and developing nations agree merely to "consideration of" certain steps. Second, developed nations agree to consideration of "nationally appropriate" commitments or actions in a way that takes explicit account of differences in circumstances. Third, developing nations agree to consideration not of "commitments" but only of "actions," and even these must be "supported and enabled" by assistance from wealthier nations. In these circumstances, it is an understatement to say that the Bali Action Plan is more symbol than substance. It is purely symbolic and, in a sense, an embarrassment for those officials and nations who sought international action to control emissions. The reason for this unhappy state of affairs is clear: Nations were acting in accordance with their short-term domestic self-interest as they perceived it.

The European Union

The European Union has shown a great deal of interest in the climate change problem, and many of its nations have contrasted their concern, and behavior, with that of the United States. And indeed, the EU has created an ambitious program, the Emissions Trading System, which covers about 40 percent of greenhouse gas emissions in the EU. Originally designed to assist the countries of the European Union in meeting their Kyoto requirements, the Emissions Trading System is a cap-and-trade system where fixed emissions allowances are allocated to industry via national allocation plans in each country.[9] An allowance must be surrendered for each ton of CO_2 emitted annually.[10] The allowances are set for multiyear periods known as trading periods; the first trading period ran from 2005 to 2007, and the second will continue through 2012.[11] Currently covered industries include energy, minerals (glass, ceramics, cement), iron and steel, and paper and pulp.[12]

The ETS has been modestly successful so far: the increase in verified emissions, adjusted for the entry of new installations, between 2005 and 2006 was only 0.3 percent, while the increase in the EU's gross domestic product was 3 percent. In 2006, 99 percent of installations also were in compliance with the regulations.[13] Another 0.68 percent emissions increase was posted in 2007.[14] Allowance transactions totaled 24.1 billion euros in 2007, a full 80 percent of the world carbon market.[15] Scholars have estimated that the ETS caused a reduction of CO_2 between 50 and 100 megatons per year, equivalent to 2.5 to 5 percent abatement.[16]

Nonetheless, the program has fallen short in many respects. During the first trading period, most countries significantly over-allocated emissions allowances—the 2005 emissions data showed a 4 percent allocation surplus—leading the price of an allowance to fall to close to zero in 2007.[17] Reallocation in 2008 for the second trading period led to an increase in prices, creating a total market disconnect.[18] Moreover, almost all of the allowances were given out freely in the first and second periods, even though researchers have suggested that auctioning allowances should be strongly preferred.[19]

Proposals made by the Commission of the European Communities for the third trading period seek to address these problems as well as to expand the overall scope of the program; they suggest that CO_2 emissions from the aviation, petrochemicals, and aluminum industries should be included in the ETS, as well as N_2O emissions from nitric, adicipic, and glycolic acid.[20] In order to reduce emissions by 20 percent in 2020, the commission recommended that allowances be decreased by 1.74 percent per year starting in 2010.[21] Moreover, it endorsed an auction system that would produce auctioning of two-thirds of the allowances by 2013.[22]

State and Local Initiatives

Climate change has been a significant political issue in the United States for many years, and while the national government has moved slowly, many state and local governments have been active.

California has led the way with several initiatives, culminating in 2006 legislation that would reduce emissions by 25 percent from business-as-usual levels. In 2003, mayors of hundreds of cities agreed to greenhouse gas caps derived from the principles of the Kyoto Protocol. In 2007, ten northeastern states agreed to reduce their greenhouse gas emissions from power plants using a cap-and-trade regime. In all, about thirty states have passed laws designed to counter climate change; many of these laws provide incentives for businesses to develop, and consumers to buy, energy-efficient products.

These laws and initiatives show that many Americans care about climate change, and politicians understand that they can enhance their popularity by promoting measures that counter this problem. However, virtually all of the activities undertaken by states and localities so far are both low-cost and wholly inadequate for addressing this problem, suggesting that politicians believe that public concern about climate change does not run deep. Even the California legislation is less significant than it might seem; legislators can, and probably will, cut back on the program if the economic costs are high. The most important point, however, is that local initiatives address the climate problem in an extremely inefficient way, and probably not at all. Raising the cost of energy production or consumption in one city or state will predictably cause people and businesses to migrate to other states, where they can continue to pollute. So if the laws have a marginal effect on the behavior of those who stay behind, the economic dislocation caused by the shifting of people and resources across state lines will be large. The optimal approach to climate change requires a uniform system of regulation across the nation, indeed across all or nearly all nations. Only national legislation enacted in a manner consistent with international commitments can accomplish this task.

Litigation

Numerous climate-related lawsuits have been launched in the United States and around the world. In other countries, plaintiffs have sought to enforce climate-related legislation. Uniquely and characteristically, plaintiffs in the United States have sought to, in effect, make climate policy.

In one set of lawsuits, plaintiffs have tried to compel the Environmental Protection Agency to bring greenhouse gas emissions under the ambit of general pollution statutes. If greenhouse gas emissions are pollutants, then the EPA must regulate certain entities that produce them. A suit eventually made its way to the Supreme Court, which held that greenhouses gases were indeed a pollutant. Under President Bush, the EPA stalled following the Supreme Court decision, but under President Obama, it has begun the process of issuing regulations.

Another set of lawsuits has included tort cases brought by individuals, public interest groups, and town and state governments, against energy producers and other corporations that emit large quantities of greenhouse gases. The complaints argue that these corporations have caused harm by emitting greenhouse gases, and that under tort principles, they should pay for their harm. For example, the town of Kivalina, Alaska, argues that Exxon and other energy producers are responsible for the melting of ice that had previously protected the town from erosion.[23] So far, these suits have had little success. Judges have expressed skepticism that they can evaluate the claims and offer remedies, given that climate change is a global problem.[24]

Whatever the legal merits of these claims, it is clear that litigation like this cannot offer a solution to the problem of climate change, and, indeed, could do more harm than good. Existing pollution statutes were not designed with climate change in mind, and thus do not give the EPA the authority that it would need to implement a sensible regulatory system—such as a tax on carbon, or a cap-and-trade system. Even worse, the tort system gives decision-making authority to the judiciary. Judges would have to cobble together a climate policy in a decentralized fashion by building up precedent, and they do not have the expertise to do this—that's why the EPA was created in the first place. The decisive problem is that even if existing tort and statutory authorities could be used to limit the emissions of the United States in substantial measure, they would have little effect on the emissions of people and corporations in foreign countries. Most advocates seem to agree that the best case for litigation is symbolic or expressive. It may bring the climate problem to the attention of the public in useful ways, but it cannot provide an adequate response to the problem of climate change. We would add that if the litigation succeeded, it would at best put in place an unnecessarily costly system of climate regulation—a high price to pay for symbolic action.

• • •

We have merely sketched the most important initiatives here; it would be easy to devote a whole volume to what has happened thus far. But the details threaten to overwhelm the key point. For all the action and energy devoted to the climate change problem, existing initiatives are mostly symbolic, not substantive. Even the most recent events in the United States, which have given some people cause for optimism, can best be described as ambiguous. The U.S. House of Representatives has passed the Waxman-Markey bill, which creates a cap-and-trade system for greenhouse gases emitted in the United States. The bill imposes real, but very low, costs on individuals (perhaps a few hundred dollars per year)[25] and businesses and, commensurately, will reduce greenhouse gas emissions by a real, but limited (about 17 percent below the 2005 level by 2020), amount.[26] The bill's main importance was to show the seriousness of the American commitment to a significant climate treaty for the next round of negotiations in Copenhagen in December 2009, but because the Senate is not expected to pass the bill by then, the world may well regard the bill as merely—symbolic.

A theme of this chapter has been that wealthy nations have been reluctant to spend substantial amounts to reduce their emissions. Kyoto was presented as a triumph of global morality, with the United States as the holdout that undermined the possibility of progress toward addressing climate change. When the treaty obligations are carefully parsed and the record of compliance is examined, it becomes clear that Kyoto and the pattern of compliance in its aftermath do not depart far from nations' perceived short-term self-interest. Countries aspired to do a great deal, agreed to do less, and accomplished very little. If the effort to put the bulk of costs on the United States was motivated by desires to do distributive or corrective justice, then the lesson of Kyoto is that there are significant limits on how far these principles can guide the conduct of states.

Chapter 4

Climate Change and Distributive Justice
Climate Change Blinders

Some nations are rich and other nations are poor. Our question in this chapter is whether rich nations have a special obligation to deal with climate change, not because they are principally responsible for the problem, but simply because they are rich. Are rich nations ethically obligated to sign a climate change agreement that is not, strictly speaking, in their self-interest, because doing so would help the poor? Shouldn't they provide disproportionate help?

Many claim that rich nations do have such an obligation. Many developing countries, for example, argue that the developed world should bear most of the cost of greenhouse gas mitigation efforts because they are rich.[1] These arguments often appeal to international-law ideas such as the right to develop, which are said to excuse developing countries from environmental and other restrictions that developed countries must observe.[2] These claims have been embodied in climate agreements. As we have seen, the Framework Convention requires that emissions reductions be based on the principle of equity and follow the principle of common but differentiated responsibilities. The Kyoto Protocol, building on the Framework Convention, imposes obligations largely on rich countries, allowing developing countries to increase their emissions without limit.[3] Some scholars have argued that rich nations should bear most or even "all of the costs" of abatement.[4]

We will argue here that these claims improperly tie valid concerns about redistribution to the problem of reducing the effects of climate change. To a great extent, these issues are and should be separate.

The rich indeed have an obligation to help the poor, but they should fulfill this obligation in the best possible way, whether this involves cash grants, development aid, trade rules, or other mechanisms. It is conceivable that climate change policies will turn out to be the best way to help poor people, but we will suggest that they are unlikely to count as such.

A serious problem is that if we choose climate policy to redistribute rather than to reduce emissions as cheaply as possible, we risk significantly raising the costs of emissions reductions or reducing their efficacy. Climate change is sufficiently serious that reducing emissions at the lowest possible cost must be our central task. As the risks from climate change increase, the problem becomes more severe: sacrificing climate change goals for distributive benefits quickly becomes a bad trade-off if failure to reduce emissions leads to terrible consequences. The greater the risk of catastrophe, the more important it is to choose the most effective climate policy. By tying the two issues together, we risk hurting both goals.[5]

This problem is particularly acute with approaches that alter when, where, and how emissions are reduced as opposed to policies that reduce emissions optimally but place the burden of paying for the reductions on rich countries. The Kyoto Protocol is a classic example of a "when, where, and how" policy. Reductions are concentrated in rich countries, significantly raising the costs of reductions. By combining distributive goals with climate policy, approaches like these risk achieving neither. Alternative approaches that ask rich countries to pay but do not change when, where, and how reductions are made, are more promising. They would not distort climate policy. Nevertheless, these approaches are questionable. There have been substantial efforts in designing aid mechanisms that help the current poor. Although these mechanisms have had only modest success, it is not clear why transfers tied to a climate treaty should replace these existing mechanisms.

We offer two claims here—one theoretical, the other empirical. The theoretical claim is that by focusing on the distributive effects of climate change, analysts miss the possibility of identifying less costly methods of helping the poor elsewhere. They operate with climate change blinders: they look only at the particular policy under consideration, here climate change, rather than considering the entire set of policies. They want each individual policy to achieve distributive goals rather than achieve redistribution through the overall set of policies. But by operating with blinders, they risk raising the costs and reducing the efficacy of both the underlying policy and the distributive goals. It is not enough to say that the rich are rich in order to impose special obligations on them with respect to a single policy such as climate change.

The empirical claim is that reducing the effects of climate change, particularly through approaches that distort climate policy, is unlikely to be a good vehicle for redistribution. It is badly targeted and expensive. For many such approaches, particularly "when, where, and how" approaches, the empirical evidence that they are costly methods for redistributive wealth is strong. Nevertheless, we agree that it is possible that the best method of redistributing to the poor will turn out to be through a climate change treaty: it might be the case that emissions can be reduced cheaply in wealthy countries and that there is no better method of transferring resources to the poor. Even if true, however, these sorts of empirical claims are not generally made and they need to be. Instead, the distributive arguments simply assume that it is enough that some countries are rich and some are poor. One of our goals here is to outline the type of empirical facts that would be necessary for climate change abatement to be a good method of redistribution.

We begin our analysis with an analogy designed to illustrate the basic point, and then turn to the specifics of climate change.

The Asteroid

Imagine that India faces a serious new threat of some kind—say, a threat of widespread devastation from the impact of an asteroid. Imagine too that the threat will not materialize for a century. Imagine finally that the threat can be substantially reduced, at a cost.

Let us stipulate that India would be devastated by having to bear that cost now; as a practical matter, it lacks the resources to do so.

But if the world acts as a whole, it can begin to build technology that will allow it to divert the asteroid, thus reducing the likelihood that it collides with India a century hence. The cost is high, but it is lower than the benefit of reducing the threat. If the world delays, it might also be able to eliminate the threat, or to reduce the damage if it comes to fruition, but most scientists believe that the best approach, considering relevant costs and benefits, is to start immediately to build technology that will divert the asteroid.

Asteroid Justice

Are wealthy nations, such the United States, obliged to contribute significant sums of money to protect India from the asteroid? It is tempting to think so. The United States is wealthy and India is poor, and when poor people need help, wealthy people should provide it. But we might initially wonder whether the help should be seen as a matter of charity or as a kind of entitlement. On one view, it is good to be charitable, and if the United States refuses to provide assistance, it is being selfish. Nations, no less than people, have a right to be selfish, but it is right to criticize selfishness.

On a quite different view, the people of India have a legitimate claim on the United States. Perhaps the obligations of some nations, and the entitlements of others, should be determined by asking about overall welfare. Perhaps nations are morally obliged to do what would maximize welfare; perhaps welfare would be higher, or greater, if the United States assisted India. After all, poor people benefit more from a dollar than rich people do, and there are many more poor people in India than in the United States. Or perhaps the obligations of nations should be determined by invoking John Rawls' idea of a "veil of ignorance": What principles of justice would reasonable people select if they were deprived of information about their own circumstances, including the nation in which they find themselves? Very plausibly, people would choose principles that would require those in rich countries to give a great deal to those in poor countries, especially if the latter are at serious risk.[6]

We think that poor people and poor nations do have a claim of entitlement, and that it is too weak to say that wealthy people and nations should be charitable to those who need help.[7] But even those who reject that conclusion are likely to agree that people should be charitable. At a minimum, we might be able to agree that the United States, and other wealthy nations, should not be selfish and should provide assistance to India if that assistance is necessary to provide protection against an asteroid collision.

The problem of the asteroid threat does have a significant difference from that of climate change, whose adverse effects are not limited to a single nation. To make the analogy closer, let us assume that all nations are threatened by the asteroid, in the sense that it is not possible to project where the collision will occur; scientists believe that each nation faces a risk. But the risk is not identical. Because of its adaptive capacity, its location, its technology, and a range of other factors, let us stipulate that the United States is less vulnerable to serious damage than (for example) India and the nations of Africa and Europe. Under plausible assumptions, the world will certainly act to divert the asteroid, and it seems clear that the United States will contribute substantial resources for that purpose. Let us suppose that all nations favor an international agreement that requires contributions to a general fund, but that because it is less vulnerable, the United States believes that the fund should be smaller than the fund favored by the more vulnerable nations of Africa and Europe and by India. From the standpoint of domestic self-interest, then, those nations with the most to lose will naturally seek a larger fund than those nations facing lower risks. Moreover, poor nations might argue for an even larger fund with the claim that spending resources this way helps the poor and vulnerable.

At first glance, it might seem intuitive to think that the United States should accept the proposal for the larger fund, simply because it is so wealthy. If resources should be redistributed from rich to poor, on the ground that redistribution would increase overall welfare or promote fairness, the intuition appears reasonable.

Difficulties

But there is an immediate problem: If redistribution from rich nations to poor nations is *generally* desirable, it is not at all clear that it should take the particular form of a deal in which the United States joins an agreement that is not in its interest. Other things being equal, the more sensible kind of redistribution would be a cash transfer, so that poor nations can use the money as they see fit. Perhaps India would prefer to spend the money that would otherwise protect potential future asteroid victims on education or on health care, for the benefit of severely impoverished people in the present. Here, then, is the initial difficulty: If redistribution is what is sought, a generous deal with respect to the threat of an asteroid collision seems a crude way of achieving it.

An asteroid treaty has some similarities to a grant of in-kind assistance, such as a grant of housing assistance to poor people, when poor people might prefer to spend the money on food or health care.[8] If redistribution is desirable, housing assistance is better than nothing, but at least initially, poor people should be given money that they can use as they see fit. It remains puzzling why wealthy nations should be willing to protect poor nations from the risks of asteroid collisions, while not being willing to give them resources with which they can set their own priorities.

There is a second difficulty. We have stipulated that the asteroid will not hit the Earth for another one hundred years. If the world takes action now, it will be spending current resources for the sake of future generations, which are likely to be much richer.[9] The current poor citizens of poor nations are probably much poorer than will be the *future* poor citizens of those nations. If the goal is to help the poor, it is odd for the United States to spend significant resources to help posterity while neglecting the present.[10] Thus far, then, the claim that the United States should join what seems, to it, to be an unjustifiably costly agreement to divert the asteroid is doubly puzzling. Poor nations would benefit more from cash transfers, and the current poor have a stronger claim to assistance than the future (less) poor.

From the standpoint of distributive justice, there is a third problem. Nations are not people; they are collections of people, ranging from very rich to very poor. Wealthy countries, such as the United States, have many poor people, and poor countries, such as India, have many rich people. If the United States is paying a lot of money to avert the threat of an asteroid collision, it would be good to know whether that cost is being paid, in turn, by wealthy Americans or by poor Americans.

If redistribution is our goal, it would also be good to know whether the beneficiaries are mostly rich or mostly poor—in our example, whether the beneficiaries are mainly wealthy Indians (suppose the asteroid is headed toward a wealthy suburb of Delhi) or poor Indians. Many of the beneficiaries of actions to reduce a worldwide risk are in wealthy nations, and so it should be clear that the class of those who are helped will include many people who are not poor at all. Because the median member of wealthy nations is wealthier than the median member of poor nations, it is plausible to think that if wealthy nations contribute a disproportionately high amount to the joint endeavor, the distributive effects will be good. For example, the Americans who are asked to make the relevant payments are, on average, wealthier than the Indians who are paying less. But asking Americans to contribute more to a joint endeavor is hardly the best way to achieve the goal of transferring wealth from the rich to the poor.

Climate Change: From Whom to Whom?

In terms of distributive justice, the problem of climate change is closely analogous to the asteroid problem. Climate change, of course, is actually caused by people, unlike the asteroid. We will discuss these issues in chapter 5. Here, we are merely dealing with the argument that rich countries should pay to reduce the likelihood of harm solely because they are rich. The asteroid, for purposes of this argument, closely resembles climate change: there is a high likelihood of serious harms in the future, harms that will impact different people differently. The question is whether concerns about distributive justice should alter who would otherwise bear the cost.

It is not always clear what people mean when they say that a climate treaty would, or should, serve distributive justice. It is sometimes argued that a climate treaty is a good way to redistribute wealth to poor countries because poor countries would gain a great deal from climate abatement measures. As we have seen, global warming will probably have a worse impact on poorer nations than on richer nations because poorer nations are in warmer areas, are more dependent on agriculture, and cannot afford seawalls and other countermeasures.

One way to state this point is that poor countries will benefit from a climate treaty, in the sense that, for them, the (discounted) gains from (future) amelioration of climate change exceed the costs of their proportionate share of abatement costs. An optimal climate treaty would be based on aggregated benefits and costs across countries. If poor states bear their share of the costs of the abatement, and enjoy the benefits of abatement, then a climate treaty is in their interest, even though it will not significantly redistribute wealth from rich to poor. On this view, a climate treaty—like a treaty to build a rocket to intercept the asteroid—benefits poor countries and therefore serves global welfare. But it also benefits rich countries and does not necessarily redistribute wealth from rich to poor.

The claim that a climate treaty should be used to redistribute wealth from rich to poor countries must mean something else, then. We distinguish three possibilities. First, the treaty imposes stricter abatement obligations—what we call "when, where, and how" policies—on rich countries than on poor countries because they are rich.[11] Second, the treaty's climate target—the reduction in emissions and hence the mitigation of climatic change—more closely tracks the interests of poor states than those of rich states; that is, rich nations might agree to reduce emissions more than they otherwise would so as to help the poor. Third, the treaty's targets and obligations are optimal from the global standpoint, but it provides for side payments from rich to poor.

When, Where, and How Policies

Claims that rich countries should bear the burden of reducing emissions are often used to support policies that change what type of emissions reductions are required—when, where, and how emissions are reduced. The Framework Convention and the Kyoto Protocol both take this approach. We have seen that the Kyoto Protocol does not impose emissions reductions on developing countries such as China, India, Indonesia, Brazil, or Malaysia, all major emitters.

Instead, reductions are concentrated in developed countries because they are wealthy.[12]

The central problem with these approaches is that they significantly raise the cost of emissions reductions and reduce their efficacy, undermining climate change goals. In addition, they may be ineffective in helping the poor compared to more direct mechanisms. To justify such approaches, advocates would have to show that concentrating emission reductions in rich countries is good climate policy (aside from the distributive effects) or that other mechanisms for transferring to the poor are sufficiently bad that it is worth adopting bad climate policy to gain the distributive benefits.

It seems relatively clear that concentrating reductions in rich countries is not good climate policy. As we saw in chapter 1, it would be almost impossible to limit carbon concentrations to a reasonable level such as 450 or 500 ppm by reducing emissions in developed countries alone. Given that the current carbon dioxide concentration in the atmosphere is about 380 ppm, that it is growing roughly 2 ppm per year, and that the world economy will probably triple in size by mid-century, emissions would need to be three times lower per unit of GDP for a 500 ppm goal to be met (and that level of reductions does not achieve stabilization of greenhouse gas concentrations). Agriculture and land use account for about 34 percent of emissions. Agriculture emissions are unlikely to fall significantly. Even if land use emissions fall somewhat, as a practical matter emissions from energy use would have to be near zero. The inevitable conclusion is that emissions reductions will have to come from developing countries, which are the first, third, fourth, fifth, and sixth biggest emitters in the world (the United States is second, China first).

Moreover, even if there were some way that emissions reductions from developed nations could in theory be sufficient, they will not be in practice. Emissions reductions only in developed nations will mean that industry from rich countries will migrate to poor countries, and the reduction in demand in rich countries will just lower the price of fossil fuels in poor countries, leading to greater consumption. Both of these offsetting effects will greatly reduce the benefits from emissions restrictions in the developed countries.

Finally, many of the low-cost abatement opportunities exist in poor nations, not rich ones. For example, it will often be cheaper to finance efficient power plants for poor nations that would otherwise build inefficient power plants on their own than to replace inefficient power plants with efficient power plants in rich nations. If the goal is to reduce warming at the lowest cost, it does not make sense for rich nations alone to scale back.

For all of these reasons, it is hard to defend the idea that wealthy nations should reduce their emissions while poor nations should be permitted to continue with business as usual.

The alternative view is that it is worth bearing the high costs in terms of bad climate policy because these approaches are the best mechanism for redistributing resources to the poor. Although we cannot rule this out given how difficult it has been to promote growth in developing countries, it seems unlikely. Recall that if wealthy countries scale back their emissions, the beneficial effects for developing countries will be modest, and hence the gains for poor people in those countries will be small. It is therefore exceedingly hard to justify, on distributive grounds, an approach that (say) requires wealthy countries to freeze their emissions at 2000 rates while asking developing countries to do nothing at all.

To be sure, we could imagine a mixed approach that might be more plausibly defended. Perhaps developed nations could be asked to freeze their emissions at 2000 rates, while developing countries could be asked to freeze their emissions at 2010 rates, or perhaps reductions from business-as-usual could vary in accordance with per capita income. An approach of this sort might well help poor nations—certainly as compared with a climate change agreement that imposed a uniform carbon tax or that required all nations to freeze their emissions as of a particular date. It is an empirical question whether such a mixed approach would be better, as an engine of redistribution, than policies that do not involve climate change at all.

Carbon Targets

All countries, or virtually all countries, gain as abatement increases. Eventually, the marginal costs equal the marginal benefits and the optimal abatement level is achieved. However, one might argue that a treaty should require a greater level of abatement than the optimal level. Such an approach would benefit poor states by mitigating climate change to a greater degree than would otherwise occur. If the poor bear the costs of these reductions, there would be no distributive effects, but if the costs are spread broadly or borne by rich nations, the additional abatement can be seen as a method of redistributing resources.

It is important to see that such a treaty would provide for a type of in-kind redistribution. The rich states bear the monetary cost of extra abatement measures, while the poor countries gain from an improved climate. The transfer is analogous to in-kind transfers to the poor of food stamps or housing assistance rather than cash. Because the world would be abating beyond the optimal amount, the costs of the additional abatement would exceed the benefits.

Such an approach has a number of problems. The treaty does not so much redistribute to poor countries today as to poor countries in the future.[13] People living in poor countries in the future will almost certainly be wealthier than people living in poor countries today. If rich countries are to bear the cost of sending wealth to future less poor people, they will spend less money on the more impoverished people of today. From the standpoint of distributive justice, this result is perverse.

Put differently, this treaty would be worse for poor countries than an optimal treaty along with cash payments to the poor countries. As we noted above, efficient policies leave more money for redistribution than inefficient policies that directly produce the distributively appropriate objective. Poor states are likely to prefer to spend more cash today on malaria nets, health care, and education, than receive a lower dollar value benefit of greenhouse gas abatement efforts in the distant future.

There is another way in which the treaty does not target the poor as well as conventional forms of foreign aid do. Most poor countries have a large class of rich people. In the future, this will still be the case. Since climate affects everyone, climatic gains will benefit the rich as well as the poor.

The optimal treaty will aim for atmospheric carbon concentrations that are optimal for the whole world. As one increases abatement, the costs and benefits increase. The benefits will increase more rapidly for some countries than for others. The first group will prefer a more aggressive treaty than the latter group will. But the optimal treaty will fall somewhere in between, disappointing both groups. If it is the case that poor countries belong to the first group, then we should regret the fact that an optimal climate treaty might have undesirable distributive effects. But we should not for that reason choose a suboptimal climate treaty, one where the marginal costs of abatement exceed the marginal benefits.

Side Payments

An alternative to Kyoto-type or other distributive policies is an efficient climate treaty—one that gets the when, where, and how of reductions right—with side payments by rich countries to help poor countries comply. At first glance, this alternative redistributive policy seems much better, and for two reasons. First, emissions reductions are, by hypothesis, implemented as inexpensively as possible. Second, money is going directly to poor nations for their current use, and in that sense the proposal might be said to help current poor people, rather than future (less) poor people. A climate change agreement in which rich nations foot a significant amount of the bill would be counted a triumph. The Montreal Protocol solved the problem of ozone depletion with an approach of this kind, which is also suggested by the Bali road map.

Side payments are cash transfers, or transfers of other valuable assets (such as tradable emission permits), that are separate from the various obligations, such as emission limits, created by the treaty. Side payments need not benefit the poor. Indeed, an optimal treaty that did not try to redistribute wealth in order to satisfy claims of justice would probably involve side payments. These side payments would not go from rich to poor, but from states that have a stronger interest in a climate treaty to states that have a weaker interest in a climate treaty. An optimal climate treaty would set the emission limits at the level that maximizes global welfare, but because some states have more favorable geographic features—they are protected from sea level rises, for example, or their climate is colder than what is desirable—they would prefer a climate treaty with looser emission limits than what are optimal for other states. To secure the participation of these states in the treaty, other states will have to make side payments to them. Because some of the favored states are wealthy, while many of the most disfavored states are poor, these side payments may go from poor to rich, at least in part.

Side payments that serve distributive justice, by contrast, would go from rich to poor, taking into account the effect of climate change over the long term. In such a scheme, rich states that would otherwise receive side payments to secure their participation might give them up and even make side payments of their own to poor states. Rich states that would not otherwise receive side payments to secure their participation will also make side payments to poor states. And poor states that would otherwise make side payments or not would receive side payments just because they are poor.

It should also be clear that side payments that are optimal from the perspective of distributive justice would not be the same thing as side payments that offset the particular costs that poor countries will face from climate abatement under an optimal treaty. Consider a middle-income country that would have very heavy burdens under an optimal treaty—perhaps because the country has a large number of dirty coal-fired power plants and extensive coal deposits that will lose much of their value. Distributive justice might counsel against providing a side payment to this country because, after all, the people who live there are not as poor as those who live in very poor countries. Meanwhile, a poor country that benefits from a climate treaty—say, it has extensive uranium deposits, need not fear sea level rises because it is landlocked, and does not depend on agriculture to a great extent—may well be entitled to side payments just because the people who live in this country will still be very poor despite the benefits of the treaty.

An optimal climate treaty with redistributive side payments is just two treaties in one—a climate treaty and what might be called a foreign aid treaty. The climate part of the treaty would provide for the optimal when, where, and how policies, along with (perhaps) side payments to secure the participation of states with weaker climate abatement interests. The foreign aid part of the treaty would send money from rich states to poor states (perhaps in the process forcing some rich states to give up side payments that they would otherwise be entitled to).

We believe that climate abatement is desirable from the standpoint of global well-being, and that foreign aid is as well. And certainly a climate treaty with optimal emissions policies, along with appropriate side payments to the poor, is superior to a climate treaty that uses suboptimal when, where, and how policies to help the poor.[14] But it is not at all clear why a climate treaty and a foreign aid treaty should be combined.

States have not so far shown any inclination to negotiate a multilateral foreign aid treaty. What reason is there to believe that they would be willing to do so in connection with a climate treaty? The answer is *not* that only foreign aid to poor states can secure their participation; as we have seen, an optimal climate treaty would probably not require side payments to poor countries. It could well require side payments to rich countries like the United States and rising countries like China, and indeed possibly from very poor countries which are extremely vulnerable to climate change—such as Bangladesh.

Certainly, there is no ethical requirement to combine a climate treaty and a foreign aid treaty into one document. The only relevance of the climate treaty for foreign aid is that if the climate treaty makes poor countries (or some poor countries) poorer and rich countries (or some rich countries) richer, then the beneficiaries will have a strengthened obligation to help those who are harmed. This obligation could be satisfied in numerous ways outside of a climate treaty—for example, simply through unilateral increases in foreign aid to affected countries.

We are concerned that if climate negotiators come to believe that a climate treaty has to serve distributive justice, two problems could arise. First, states would be unable to agree for the same reason that states have never agreed on a multilateral foreign aid treaty—there are just too many conflicting views about who deserves aid and who should pay it. This is why most foreign aid is unilateral. It would be terrible if states fail to agree on a climate treaty with optimal when, where, and how policies because they cannot agree on the magnitude and allocation of foreign aid, and who should be most responsible for it.

Second, we suspect that states would not be able to give as much thought to the foreign aid portion of a climate treaty as they should. Climate science and economics are complicated enough, and most of the effort will be devoted to mapping out how obligations should turn on the scientific and economic uncertainties. The allocation of foreign aid would be at best an afterthought.

Foreign aid is an extremely difficult problem, one that is in its own way every bit as complicated as climate science. Nations have experimented with foreign aid for many decades and have little to show for the effort. Although the effectiveness of foreign aid is still a matter of debate, everyone agrees that foreign aid programs work only if they are extremely carefully designed and carried out.[15] Yet in climate treaty discussions, none of these issues have been addressed. The question of distribution is not treated as a question of foreign aid, but as a question of fairness. Poorer countries will receive more permits because they are poor, and the fact that poor countries have rarely benefited from distribution of cash or in-kind foreign aid receives no attention.

It is easy to understand why. Poor countries would resent the paternalistic assumptions behind such an approach to the distribution of permits. They would almost certainly reject the conditions that are routinely used for foreign aid—inspection, monitoring, reporting, and so forth. Rich countries, if committed to the idea that justice requires distributions of permits to poor countries, would have to yield. And yet if experiences with foreign aid are any evidence, the massive distribution of wealth to poor countries would be squandered.

Why would distributive justice demand such an outcome? Let rich nations continue to experiment with foreign aid programs for poor countries, and meanwhile allow all states to negotiate a climate treaty that satisfies International Paretianism but that does not try to redistribute to the poor.

If these arguments are right, then distributional justice does not have much to say about the design of a climate treaty. The treaty should stipulate emissions that are optimal for the globe. Such a treaty could end up stipulating a level of abatement that falls short of what poor nations prefer. There is an entirely separate question of what the rich nations owe to the poor nations, and the answer to this question should determine aid flows and other projects. Of course, it is theoretically possible that the best way to transfer from rich to poor is through abatement in excess of what the rich world prefers, but as we have argued, this theoretical possibility is just that. The best way to help poor people today is unlikely to be to undergo projects that will have no significant effect for many decades. A climate change treaty should focus on climate change, not on other goals.

Rescuing Intuitions? Four Counterarguments

There are four tempting counterarguments to our criticisms of a redistributive climate treaty. The first involves the risk of catastrophe. The second involves the fact that cash transfers will go to governments that may be ineffective or corrupt, while reducing the effects of climate change affects people directly. The third is that delayed commitments to reduce emissions by developing countries are to be preferred for a variety of reasons. The fourth is that redistribution through a climate change treaty is feasible while direct redistribution is not; the world's attention, for example, might be focused on the poor while negotiating a climate treaty.

Catastrophe

On certain assumptions about the science, greenhouse gas reductions are necessary to prevent a catastrophic loss of life. The risk of catastrophe, even if small, may drive policy decisions.[16] Suppose, as is not entirely unrealistic, that business-as-usual creates a small risk of increasing global mean temperatures by 10°C by 2100. A change of this kind would fundamentally change human life and risk many millions of deaths. How does this point affect the analysis of distributive justice?

If other nations could eliminate the harms associated with such a disaster, it would be hard to object that they should offer cash payments instead. One reason is that if many lives are at risk, and if they can be saved through identifiable steps, taking those steps would seem to be the most effective response to the problem, and cash transfers would have little or no advantage. The cash would be best spent on eliminating the harm in any event. If poor people in poor nations face a serious risk of catastrophe, then greenhouse gas abatement could turn out to be the best way to redistribute to people who would otherwise die in the future.

The argument works, however, only for what we have called efficient policies with side payments. "When, where and how" policies look far worse in the case of catastrophe because they sacrifice climate change goals for redistribution. If catastrophe is a serious possibility, the last thing we should do is to sacrifice emissions reductions. Doing so may very well make those we are trying to help worse off. For example, Kyoto will not be as effective in reducing greenhouse gases as a treaty that included all sources of emissions from all countries. If a poor country faces the risk of catastrophe from climate change, excusing it from emissions reductions may not be helpful: what it gains in distributive benefits in the climate treaty is taken back in the form of risk of catastrophe.

The catastrophe argument offers important considerations, and may change the conclusions, but it does not fundamentally change the basic analysis. It has three interpretations. First, it is a way of saying that the welfare gains from a climate treaty are very high. If so, then obviously a higher level of abatement is justified, but there are no particular implications for distributive justice. Second, it is a way of saying that foreign aid turns out to be the same as or inferior to abatement as a means of redistributing wealth. As we noted earlier, if this is the case, then from the standpoint of distributive justice, rich nations better satisfy their goals if they agree to (from their internal view) excessive abatement measures than if they distribute foreign aid. Third, as the possibility of catastrophe gets higher, it becomes all the more important to avoid policies that distort when, where, and how emissions are reduced.

Ineffective or Corrupt Governments

We have emphasized that in principle, development aid is likely to be more effective than greenhouse gas restrictions as a method of helping poor people in poor nations. But there are notorious difficulties in making development aid effective. Even when rich countries want to help, it is hard for them to do so. A legitimate argument for cutting greenhouse gas emissions, or for ensuring that financial assistance goes directly to ensure emissions reductions, is that these steps bypass the governments of poor states more completely than other forms of development aid do. This might be counted as a virtue because the governments of poor states are, to a large degree, either inefficient or corrupt (or both), and partly for that reason, ordinary development aid has not been very effective.[17] Emissions abatement might be regarded as worse, in principle, than effective cash transfers, but if effective transfers are not feasible, emissions abatement might be seen as a defensible "second-best."

On the other hand, this form of redistribution does not, as we have stressed, primarily help existing poor people; it largely helps poor people in future generations. And it is not clear that donor states can avoid the various problems of development aid by, in effect, transferring resources to the future rather than to the present, or by transferring resources directly to the people rather than to corrupt governments. If the corrupt governments of today can borrow against the future, or raise additional revenues by increasing taxation today, then the effectiveness of these strategies will be diminished.

Indeed, donor groups have long understood the problem of corrupt and inept governments and have tried to circumvent them in many ways. For example, donors give aid directly to farmers, businesses, private charitable organizations, medical clinics, and so forth. They provide training and other in-kind resources to government employees. They try to identify and work with government reformers who are competent and honest. Unfortunately, none of these strategies has worked terribly well.[18] So, at the least, the argument that redistribution through greenhouse gas reduction will work where these other approaches have failed needs to identify the distinctive characteristics of greenhouse gas reduction that provide grounds for optimism. This argument would also need to take account of the major disadvantage of greenhouse gas reduction compared to the other redistributive instruments: it does not target people who are especially needy or take advantage of local knowledge about the particular problems faced by the recipient state.

Even more important, the claim that emissions reductions avoid corruption overlooks the fact that emissions abatement does not occur by itself but must take place through the activity of governments. In cap-and-trade systems, for example, the government of a poor country would be given permits, which it could then sell to industry, raising enormous sums of money that the government could spend however it chose. Corrupt governments would spend this money badly, perhaps using it to finance political repression, while also possibly accepting bribes from local industry that chooses not to buy permits, in return for non-enforcement of the country's treaty obligations.[19] Even the much simpler carbon tax approach would not avoid these problems. Corrupt governments could simply refrain from levying the tax on cronies or offsetting the taxes with hidden transfers. Policing even the simplest climate treaty presents a serious challenge, and the stricter the obligations that the treaty imposes, the stronger the incentives to violate them, in which case policing would be that much more difficult.

The point for present purposes is that in principle, greenhouse gas cuts do not seem to be the most direct or effective means of helping poor people or poor nations. We cannot exclude the possibility that the more direct methods are inferior, for example because it is not feasible to provide that direct aid, but it would remain necessary to explain why a crude form of redistribution is feasible when a less crude form is not.

Delayed Reductions by Developing Countries

As a practical matter, an effective climate treaty is likely to impose obligations on developed countries to reduce emissions sooner than developing countries. We might see, for example, a treaty in which Annex I countries agree to start reducing emissions immediately while other countries agree to reduce emissions starting in ten years or some other intermediate time period. Such an approach, it might be argued, is mandated by distributive concerns because it avoids imposing crushing burdens on developing nations while at the same time eventually ensuring global emissions reductions.

While we agree that distributive concerns might justify such an approach—we mentioned this above—the strong intuitions behind this approach are likely driven by other concerns. In particular, this approach might be justified entirely separately from distributive concerns; it might very well be the most efficient or optimal policy. The reason is that reducing emissions requires strong institutions; governments must be able to monitor emissions and impose limits on powerful industries. It is possible that many poor nations would need a substantial start-up period to develop this capacity. That is, it is possible that the lowest cost reductions initially occur in wealthy countries, so emissions reductions should start in these countries entirely without regard to the distributive effects. The issue would simply be one of feasibility.

However, it is also possible that such an approach might be a serious mistake. Many low-cost reductions are likely to be found in developing countries. Moreover, decisions about the types of power plants, building construction methods, the location of housing, and transportation networks have long-lasting effects. Delaying emissions reductions obligations by fast-developing nations may mean that these nations lock in bad infrastructure decisions, making long-term emission reductions more expensive. Distributive justice arguments supporting this approach to a climate treaty in this event should be rejected.

The World's Attention

A final argument for using climate change abatement as a means of redistribution is that because of the obvious distributive effects, a climate change treaty is an opportunity to redistribute that is not otherwise available. To the extent that redistribution now is not sufficient, a climate change treaty might be an opportunity to go further.

Although we cannot refute this empirical speculation as an abstract matter, we see no reason why it should be so. There is substantial attention on rich/poor or north/south differences. For example, celebrities of various kinds travel the world arguing for debt relief; access to medical care such as AIDS vaccines and the like are a constant source of discussion. Although we agree that rich countries should do more, it is not clear why combining the already enormously difficult issue of climate change with the enormously difficult issue of redistribution would make either one easier.

A separate claim, which we do accept, is that to the extent climate change makes the world less equal, rich countries have greater duties to the poor. There is already a large spread between rich and poor. Climate change threatens to increase this difference. Therefore, climate change may increase the obligations of rich countries. This does not, however, say anything about how best to fulfill those obligations. It may turn out that reducing emissions is the best method of fulfilling these obligations, but it very likely will not be.

A More Limited Role for Redistributive Concerns?

One can distinguish more and less aggressive arguments about redistribution. Our argument in this chapter so far has been directed at the most aggressive argument: namely, that a climate treaty would be a desirable instrument for redistributing wealth from rich countries to poor countries. In the extreme, this argument suggests that rich countries should agree to a climate treaty that produces net costs for them—that is, violates the condition of International Paretianism. However, more limited claims could also be made.

Consider, for example, the following modestly redistributive treaty. Suppose that rich state R will lose 80 from global warming, and poor state P will lose 50. A climate pact will reduce these costs to 0, but would itself involve abatement costs of 100. These abatement costs could be allocated to either of the states, or both, in any proportion. For example, if the treaty creates a cap-and-trade system, the permits could be allocated entirely to R, entirely to P, or divided between them.

Clearly, a treaty would be desirable: global costs are 100, while global benefits are 130. International Paretianism requires that state R bear no more than 80 of the costs, while state P bears no more than 50. Otherwise, neither state would find it in its narrow interest to enter the treaty.

Suppose that R is fantastically richer than P—just as the United States is fantastically richer than Malawi. Shouldn't the United States be required to bear more than 80? Perhaps R should bear all the costs—100—for the sake of helping P? Indeed, perhaps R should agree to an even more aggressive and highly costly treaty for P, one that would (say) cost R another 100 even though the benefits for P would be only 20?

As we have suggested, there is a question of feasibility: whether R simply would not consent to a treaty that makes it worse off. And even if citizens of R feel some degree of altruism toward poor people in P, there are most likely better ways of helping these people than through an aggressive climate treaty.

But now consider the more modest argument. As long as the climate treaty satisfies International Paretianism, the "surplus"—here, 30—should be given entirely to poor countries. R should not complain: the treaty makes it better off. And using its immense bargaining power to demand all or a large share of the surplus would offend distributive justice.

State R loses 80 from climate change; thus, if it pays 80 for abatement, it would be just indifferent between a treaty and no treaty. If R pays 80, then P pays 20. So R gains 0, while P gains 30. At the opposite extreme, State P loses 50 from climate change; thus, if it pays 50 for abatement, it would be indifferent between a treaty and no treaty. If P pays 50, then R pays 50. So R gains 30, while P gains zero. The parties could also divide the surplus: R pays 65 and P pays 35; R gains 15 and P gains 15. Other combinations are possible.

How should the surplus be divided? Simple fairness might argue in favor of an equal division. But distributive justice would argue in favor of the first option: R pays 80 and P pays 20, so that R gains

0 and P gains 30. An alternative possibility is that ethical concerns should play no role in the division of the surplus itself, and just that relative bargaining power will do so.

A case can be made for the first option on distributive justice grounds, but it is subject to numerous qualifications. First, many people might think that a fair (equal) division is more appropriate. Although we do not share this view, it is not easy to show why it is wrong.

Second, one needs to keep an eye out for perverse consequences of such a deal. Rich states already give foreign aid to poor states. If this level of aid reflects the limits of the public's altruism, then a climate deal that effectively enhances the level of foreign aid would likely lead the original level of direct aid to be reduced. In addition, treaties that reward states for being poor will encourage them to stay that way—rather than building institutions and making the political sacrifices that are necessary for economic growth. We will return to this point in chapter 8.

Third, the effectiveness and honesty of the government of P are relevant considerations. If the government is likely to squander its disproportionate share of the surplus, or distribute it to rich elites rather than poor masses, then there are no ethical obligations on the part of R to refrain from using its bargaining power to obtain the largest share of the surplus possible.

Fourth, and most important, it is far from clear that R can do best from the standpoint of distributive justice by, in effect, over-paying P for its cooperation. Consider this analogy. A poor street vendor sells sunglasses for $20 each. A rich person would like to buy the sunglasses, and $20 would just compensate the vendor for his costs. Does distributive justice require the rich buyer to pay more than $20 for the sunglasses—$30, or $300, or $3,000? It seems that if the rich person cares to advance distributive justice, he should pay $20 and donate a substantial sum to a reputable charitable organization that helps the truly poor, rather than overpay the street vendor whom he happens to encounter. Similarly, it would seem that distributive justice has little to say about R and P's division of the cost of the climate abatement. It may be that R has an independent obligation to help even poorer countries, or that its obligation to help this particular P is greater or less than the 15 or 30 at stake in the negotiations.

The upshot is that while distributing the surplus in favor of poor countries satisfies International Paretianism and hence cannot be ruled out on feasibility grounds, this approach deserves skepticism. It may lead to perverse incentives and not serve justice in a rational and effective way.

Distributive Justice and Climate Change: Taking Stock

Rich countries have complex foreign policies and, in practice, different means often serve diverse goals in ways that sometimes obscure what those goals are. But it is important to maintain conceptual clarity. Among its many foreign policy goals, the United States, like other wealthy nations, sometimes seeks to help poor people in poor nations, and it should also seek to address the problem of climate change. The problem of poverty is one of distributive justice, and the principles of distributive justice should guide one's approach to it. The rich should contribute to the poor where the contributions improve their well-being. The poorest and most desperately in need should have priority; but when aid is ineffective because of corruption, or the difficulty of understanding foreign cultures, or some other reason, then it may well be inappropriate. Finally, aid should take the best form possible. Presumptively, cash aid is best because it can be used by recipients in a manner that they believe best advances their interests; but sometimes in-kind aid will have advantages.

The climate change problem is the failure of international cooperation to create a public good. Environmental harm in the future can be mitigated through cooperation today. Cooperation involves costly action by diverse nations that benefit to a different extent from greenhouse gas reductions. Because of this diversity, a successful climate treaty will undoubtedly put greater burdens on some states than on others. States that believe they will be harmed relatively less by climate change are likely to hold out, and either insist on limited abatement measures or agree to greater measures only if they receive side payments. These need not be the poorest nations; they may well be rich nations, and they are most likely to include moderately well-off developing giants like China and Brazil, which have been highly resistant to emissions restrictions. Thus, there is almost certainly a conflict between the requirements of distributive justice and the kind of treaty that is likely to be ratified.

It is important to distinguish between globally optimal abatement and the distribution of abatement costs. The treaty should stipulate the globally optimal abatement even though that is not what is optimal for poor states. As we have explained, transferring wealth to poor states through changes to climate policies is an extremely crude way to redistribute wealth. As for the distribution of the costs, it makes no difference from the standpoint of distributive justice if rich states transfer money to poor states by giving them foreign aid directly or by giving them foreign aid in connection with a climate (or any other) treaty, calling that aid "emission permits" or just cash. The point is that redistribution should be to the poorest states such as Bhutan, not the states that are hit hardest by the climate treaty, many of which will be middle-income or moderately poor (see chapter 1). From the perspective of distributive justice, the relevant question is not who is hurt more or less by climate change, or by a cutback in industrial activity caused by the implicit tax on energy, but who is in the worst shape overall, whether that state's problems are linked to climate change or not.

It is possible that some states with a strong bargaining position and an interest in limited abatement measures could agree to more aggressive measures because their redistributive goals are better served by a generous climate treaty than by cash transfers and other forms of foreign aid. But we are skeptical that this possibility could play more than a modest role in the design of a treaty.

The important point to remember is that if redistribution is the goal, one should make comparisons of the different ways of redistributing wealth and choose the best. In the climate debate, as in many other debates about international policy, it is instead simply assumed that the rich should pay more because they are rich.

There is an irony, and even a risk of tragedy, in the background here. Those who care about distributive justice could very well subvert a climate treaty if they insist that the wealth of a state should determine its burdens. A narrow treaty that serves the interest of states is much more likely to be ratified and to survive pressures to cheat as time passes. Here as elsewhere, arguments from principle are likely to run into the practical realities of the state system.

Chapter 5

Punishing the Wrongdoers
A Climate Guilt Clause?

In the last chapter, we imagined an asteroid hurtling toward earth, and asked how the burdens of intercepting and destroying the asteroid should be shared among the nations of the world. Many people would object that the asteroid example is misleading because it lacks a characteristic that is fundamental to the climate change problem. The asteroid is nobody's fault, it is argued, while climate change is the fault of the rich, industrial nations, which have contributed greenhouse gases to the atmosphere out of proportion to their population or their needs. The United States, for example, has 300 million people, but in 2005 contributed 18.4 percent of the carbon dioxide emissions (excluding land use change). By contrast, in that year India had over 1 billion people and contributed only 5 percent of the carbon dioxide emissions. If one looks at stocks rather than flows, the story is the same. Between 1950 and 2000, the United States contributed 17 percent of the current stock of greenhouse gas emissions; India has contributed well under 2 percent.

The upshot is a very ugly picture that depicts the citizens of wealthy countries, including the European countries, Canada, Japan, and Australia, as well as the United States, consuming wasteful goods such as SUVs and heated swimming pools over many decades, while people in the poorest countries have barely had enough to eat. Finally, after many decades of poverty, some developing nations are set to deliver reasonably comfortable standards of living to their citizens—although still far short of what prevails in the West—and then are told that they are going to have to pay a large share of abatement costs to address a problem that is the result of the profligacy of the West. These abatement costs will come out of the pockets of the citizens of those still poor countries, in the form of taxes and higher prices for consumer goods. Wouldn't it be fairer for the wealthy countries to incur all or most of the abatement costs, while the poorer countries continue to catch up?[1]

The argument that rich countries should pay for climate abatement, or pay for most of climate abatement, because they are most responsible for the problem of climate change is an argument about what philosophers call *corrective justice*. According to corrective justice, if one person harms another person, the first person should provide a remedy, such as cash, to the victim. Applying this idea to the setting of climate change, developing countries and their supporters argue that because the United States and wealthy countries have caused, or mostly caused, the problem of climate change, they should provide the remedy. Some people have argued that these rich countries should literally pay compensation to people who are now suffering the ill effects of global warming, such as people who live in low-lying villages that are being flooded with greater frequency. But the usual argument is that the rich countries should pay most of the cost of greenhouse gas abatement—that is, the rich countries alone should be forced to limit their emissions, or (more plausibly) the rich countries should have relatively fewer greenhouse gas emission permits, so that they will have to pay poorer countries for their permits in order to continue to pollute.[2]

The idea that the people who pollute should pay for the harm they cause to the environment is often called the "polluter pays principle." The polluter pays principle can be derived from the corrective justice argument and is often identified with it, but it can be given other normative justifications as well.

The corrective justice argument appeals to some powerful intuitions. However, we will argue in this chapter that it has serious flaws, and does not provide useful guidance for the design of a climate treaty. As we have seen in chapter 1, the argument encounters serious problems on the facts. As of 2005, for example, China accounted for 10 percent of cumulative emissions,[3] and the developing countries were not far behind the developed ones in terms of total emissions. Recall too that explosive emissions growth is occurring in the developing world. By 2030, it will be hard to argue, on the facts, that the developed nations are largely responsible for the total harm. In addition, if one defines harm on a per-person basis rather than on a per-country basis, the richest nations are no longer the worst offenders—instead, some very poor nations are. We will return to these issues in due course.

It is true, however, that emissions in some countries have imposed serious risks on others, that the United States and China are expected to remain the world's leading contributors, and that some nations, including those in Africa, face serious risks even though their own emissions are trivial. India's emissions are hardly trivial, but that nation might also claim that it faces serious risks for which it is not responsible. Even so, the corrective justice argument faces serious difficulties. The reason, briefly, is that the argument, in the form described above, assumes that nation-states are the relevant moral agents, and that when one nation-state injures another nation-state, corrective justice requires that the first owes a remedy to the second. But the idea that nation-states can be moral agents is highly unappealing, as it relies on notions of collective responsibility that have been rejected by mainstream philosophers as well as institutions such as criminal and tort law. Collective responsibility implies that if one person in a group (say, the father in a family, or a soldier in a nation-state) commits a wrong against another person, then the victim has a remedy against other members of the group (the father's child, the soldier's co-citizens). Collective responsibility once played an important role in social organization, but over the centuries, it has been progressively squeezed out of domestic and international law. Although it sometimes has instrumental value,[4] no one seems to defend it anymore as a matter of principle.

The corrective justice argument for putting the climate burden on the leading emitters can be resurrected in a morally appealing form, one that does not rely on notions of collective responsibility and instead makes the standard assumption of individualism. On this approach, we would need to identify particular individuals who, through their activities (for example, driving), have caused damage to the climate that has harmed other individuals. In principle, we would allow the victims some type of remedy against the wrongdoers or, if this is difficult, we might permit a form of rough justice that, in some aggregate sense, ensures some kind of transfer from a sufficiently large fraction of the wrongdoers to a sufficiently large fraction of the victims. Although we cannot rule out such an approach, we will argue that it is also not appealing. The problem is that, on any realistic assessment of wrongdoing, the number of culpable contributors to the climate problem who are alive today is a modest fraction of all contributors, and the number of victims today (as opposed to in the future) is quite small; indeed, it may make little sense to say that any person is individually culpable, given the nature of the problem.

The Basic Argument

Corrective justice arguments are backward-looking, focused on wrongful behavior that occurred in the past.[5] Corrective justice therefore requires us to look at stocks rather than flows. In the context of climate change, the corrective justice argument is that some nations have wrongfully harmed the rest of the world—especially low-lying states and others that are most vulnerable to global warming—by emitting greenhouse gases in vast quantities. On a widespread view, corrective justice requires that those nations devote significant resources to remedying the problem[6]—perhaps by paying damages, agreeing to extensive emissions reductions, or participating in a climate pact that is not in their self-interest. India, for example, might be thought to have a moral claim against the United States—one derived from the principles of corrective justice—and on this view the United States has an obligation to provide a compensatory remedy to India. (Because the United States is the leading contributor to the existing stock of emissions, we use that nation as a placeholder for those who have inflicted harms on others; and because India is especially vulnerable to climate change, we use that nation as a placeholder for those at particular risk.)

This argument enjoys a great deal of support, and seems intuitively correct. We shall identify a large number of problems here, and the discussion will be lamentably complex. The most general point, summarizing the argument as a whole, is that the climate change problem poorly fits the corrective justice model because the consequence of tort-like thinking would be to force many people who have not acted wrongfully to provide a remedy to many people who have not been victimized. Some of the problems we identify could be reduced if it were possible to trace complex causal chains with great precision; unfortunately, this will not be possible.

The Wrongdoer Identity Problem

Corrective justice normally requires an injurer to compensate a victim for harm the injurer caused. For such an obligation to arise, we must be able to identify an injurer who behaved in a morally culpable way. We cannot simply label all Americans or all individuals who live in other developed nations as morally culpable injurers. Americans vary dramatically with respect to their individual emissions and whether they have behaved culpably.

Consider table 5.1, which shows the percentages of the current American stock of emissions that are the result of emissions since a particular year. According to the U.S. Census Bureau, 54.5 percent of Americans were born after 1975.[7] These individuals are not responsible for the more than half of the U.S. emissions that occurred prior to this time. More than 27 percent of Americans are currently younger than twenty years old and are arguably not responsible at all for emissions—they do not get to choose where to live or what size house to buy.

As we will discuss later in this chapter, there is a further question about whether the people who have caused these emissions are morally responsible for them. Normally, moral responsibility requires culpability as well as causation. Yet many Americans today do not support the current American energy policy and already make some sacrifices to reduce the greenhouse gas emissions that result from their behavior. They avoid driving, they turn down the heat in their

Table 5.1.
Percentage of American Stock of Emissions by Quarter Century[a]

Since	Megatons of CO_2	% of Stock
1850	342,603	100
1875	323,040	94
1900	314,772	92
1925	282,967	83
1950	236,035	69
1975	152,593	45
2000	28,988	8

[a] Calculations from Climate Analysis Indicator Tool [CAIT], available at http://www.cait.wri.org. The data represent the sum of emissions for the relevant time periods without accounting for the decay of greenhouse gases in the atmosphere This is the same methodology used for computing stocks in chapter 1.

homes, and they support electoral candidates who advocate greener policies. Holding these people responsible for the wrongful activities of people who lived in the past seems perverse.

To be sure, many Americans have made choices that do not adequately take climate consequences into account. But even among these Americans, we should distinguish between greenhouse gas emissions that occurred before the problem of anthropogenic climate change was widely known, or before that point at which reasonable people would have acknowledged the problem, and later emissions. When was that point? In the 1990s? In the 2000s? It can take quite a while for a scientific consensus to filter down to the general public. If so, the percentage of potentially "culpable" emissions in today's U.S.-generated stock is down to a relatively small number.

Finally, there is a question of whether people are morally obliged to cut back on greenhouse gas emissions if others are not doing the same. This is a "moral collective action problem"—a problem to which we will return. For present purposes, the main point is that corrective justice usually distinguishes between people who are at least causally responsible for the problem (many of whom are dead), and more likely, those who are culpable, on the one hand, and those who are neither causally responsible nor culpable, on the other. "America" is not culpable; certain Americans may be. An approach that emphasized corrective justice would attempt to focus on particular actors, rather than Americans as a class, which would appear to violate deeply held moral objections to collective responsibility.[8]

A natural response to this point is to insist that all or most Americans today benefit from the greenhouse gas emitting activities of Americans living in the past, and therefore it would not be wrong to require Americans today to pay for abatement measures. This argument is familiar from debates about slave reparations, where it is argued that Americans today have benefited from the toil of slaves 150 years ago.[9] To the extent that members of current generations have gained from past wrongdoing, it may well make sense to ask them to compensate or make whole those harmed as a result. On one view, compensation can work to restore the status quo ante, that is, to put members of different groups, and citizens of different nations, in the position that they would have occupied if the wrongdoing had not occurred.

In the context of climate change, however, this argument runs into serious problems. The most obvious difficulty is empirical. It is true that many Americans benefit from past greenhouse gas emissions, but how many benefit, and how much do they benefit? Many Americans today are, of course, immigrants or children of immigrants, and so not the descendants of greenhouse gas emitting Americans of the past. Such people may nonetheless gain from past emissions because they enjoy the kind of technological advance and material wealth that those emissions made possible. But have they actually benefited, and to what degree? Not all Americans inherit the wealth of their ancestors, and even those who do would not necessarily have inherited less if their ancestors' generations had not engaged in the greenhouse gas emitting activities. The idea of corrective justice, building on the tort analogy, does not seem to fit the climate change situation.

Suppose that these various obstacles could be overcome and that we could trace, with sufficient accuracy, the extent to which current Americans have benefited from past emissions. As long as the costs are being toted up, the benefits should be as well and used to offset the requirements of corrective justice. If past generations of Americans have imposed costs on the rest of the world, they have also conferred substantial benefits. American industrial activity has produced products that were consumed in foreign countries, for example, and has driven technological advances from which citizens in other countries have gained. Many of these benefits are positive externalities, for which Americans have not been fully compensated. To be sure, many citizens in, say, India have not much benefited from those advances, just as many citizens of the United States have not much benefited from them. But what would the world, or India, look like if the United States had engaged in 10 percent of its level of greenhouse gas emissions, or 20 percent, or 40 percent? For purposes of corrective justice, a proper accounting would seem to be necessary, and it presents formidable empirical and conceptual problems.

In the context of slave reparations, the analogous points have led to interminable debates, again empirical and conceptual, about historical causation and difficult counterfactuals.[10] But-for causation arguments, used in standard legal analysis and conventional for purposes of corrective justice, present serious and perhaps insuperable problems when applied historically. We can meaningfully ask whether an accident would have occurred if the driver had operated the vehicle more carefully, but conceptual and empirical questions make it difficult to answer the question whether and to what extent white Americans today would have been worse off if there had been no slavery—and difficult too to ask whether Indians would be better off today if Americans of prior generations had not emitted greenhouse gases. In this hypothetical world of limited industrialization in the United States, India would be an entirely different country, and the rest of the world would be unrecognizably different as well.

The Analogy to Corporate Liability

Proponents of slave reparations have sometimes appealed to principles of corporate liability. Corporations can be immortal, and many corporations today benefited from the slave economy in the nineteenth century. Corporations are collectivities, not individuals, yet they can be held liable for their actions, which means that shareholders today are "punished" (in the sense of losing share value) as a result of actions taken by managers and employees long before the shareholders obtained their ownership interest. Indeed, lawsuits against corporations and other entities that benefited from slavery and other wrongdoing from the distant past have met with some success. If innocent shareholders can be made to pay for the wrongdoing of employees who are long gone, why can't citizens be made to pay for the wrongful actions of citizens who lived in the past?

The best answer is that corporate liability is most easily justified on grounds other than corrective justice. Shareholder liability can be defended on the basis of consent or (in our view most plausibly) on the welfarist ground that corporate liability deters employees from engaging in wrongdoing on behalf of the corporate entity. A factor that distinguishes shareholder liability based on corporate wrongdoing from liability based on citizenship in a culpable state is that purchasing shares is a voluntary activity and one does so with the knowledge that the share price will decline if a past legal violation comes to light, and this is reflected in the share price at the time of purchase. (One also benefits if an unknown past action enhances the value of the company.) But because the corporate form itself is a fiction, and the shareholders today are different from the wrongdoers yesterday, corporate liability cannot be grounded in corrective justice.[11]

Welfarist or related efficiency-based justifications for corporate liability can also be given. If corporations are made liable for the wrongdoing of employees, then shareholders are given an incentive to encourage managers to monitor and discipline employees. Shareholders cannot escape liability merely by selling their shares to third parties, because third parties will pay less for shares if they believe that managers have not sufficiently controlled employees. Such a welfarist argument can also be applied to states, albeit in a highly modified form—given the problem that citizens don't have as much power over governments as shareholders have over management, and cannot as easily exit—and that is what we will do in chapter 8. For now, it is sufficient to observe that the welfarist argument and corrective justice argument are not the same, and the practice of corporate liability does not provide an argument in favor of applying corrective justice arguments to states in the setting of climate change treaty negotiations.

The Victim/Claimant Identity Problem

As usually understood, corrective justice requires an identity between the victim and the claimant: the person who is injured by the wrongdoer must be the same as the person who has a claim against the wrongdoer. In limited circumstances, a child or other dependent might inherit that claim, but usually one thinks of the dependent as having a separate claim, deriving from the wrongdoer's presumed knowledge that by harming the victim she also harms the victim's dependents.

Who are the victims of climate change? Most of them live in the future. Thus, their claims have not matured. To say that future Indians might have a valid claim against Americans today, or Americans of the past, is not the same as saying that Americans today have a duty to help Indians today. To be sure, some people are now harmed by climate change. In addition, people living in low-lying islands or coastal regions can plausibly contend that a particular flood or storm has some probabilistic relationship with climate change—but from the standpoint of corrective justice, this group presents its own difficulties (a point to which we will return shortly). What remains plausible is the claim that future Indians would have corrective justice claims against current and past Americans.

A successful abatement program would, of course, benefit many people living in the future, albeit by preventing them from becoming victims in the first place or reducing the magnitude of their injury, rather than compensating them for harm. Such an argument does not rely on principles of corrective justice—it is forward-looking rather than backward-looking. It does reflect a concern about avoiding harm. But there is no deontological restriction on engaging in behavior that imposes risks on people in the future—virtually all behavior imposes risks on others, and some other principle is needed to distinguish between risky behaviors that are justified and those that are not. Governments generally take a welfarist approach to such behavior, banning or restricting activities that impose high risks of harm and permitting other activities.

One might justify the abatement approach on welfarist grounds: perhaps the welfare benefits for people living in the future exceed the welfare losses to people living today. That is the view that we defend in chapter 8. One could also make an argument that people living today have a nonwelfarist obligation to refrain from engaging in actions today that harm people in the future. The point for present purposes is that both arguments are forward-looking: the obligation, whether welfarist or nonwelfarist, is not based on past actions, and thus a nation's relative contribution to the current greenhouse gas stock in the atmosphere would not be a relevant consideration in the design of the greenhouse gas abatement program, as we have been arguing. By contrast, the corrective justice argument is that the United States should contribute the most to abatement efforts because it has caused the most damage to the carbon-absorbing capacity of the atmosphere.[12]

The Causation Problem

Corrective justice requires that the wrongdoing cause the harm. In ordinary person-to-person encounters, this requirement is straightforward. But in the context of climate change, causation poses significant challenges, especially when we are trying to attribute particular losses to a warmer climate.

To see why, consider a village in Alaska that was formerly protected from erosion by ice that blocked ocean waves but now must build expensive barriers or even relocate because the ice has melted. One might make a plausible argument that the ice would not have melted but for global warming. But it might well be impossible to show that greenhouse gas emissions in the United States "caused" the melting of the ice and the increased erosion, in the sense that they were a necessary and sufficient condition, and it would be difficult to show that they even contributed to it. If the erosion was in a probabilistic sense the result of greenhouse gas activities around the world, its likelihood was also increased by complex natural phenomena that are poorly understood.

Causation problems are not fatal to corrective justice claims, but they significantly weaken them. In tort law, courts are occasionally willing to assign liability according to market share when multiple firms contribute to a harm—for example, pollution or dangerous products whose provenance cannot be traced.[13] And it would be plausible to understand corrective justice, in this domain, in probabilistic terms, with the thought that victims should receive "probabilistic recoveries," understood as the fraction of their injury that is probabilistically connected with climate change. It is unclear, however, that statistical relationships can be established with sufficient clarity to support a claim sounding in corrective justice.[14]

The Culpability Problem

Philosophers disagree about whether corrective justice requires culpability.[15] Intentional, reckless, or negligent action is usually thought to be required for a corrective justice claim. While some people do support strict liability on corrective justice grounds, a degree of culpability is required to make the analysis tractable. Because multiple persons and actions (including those of the victim) are necessary for harm to have occurred, identification of the person who has "caused" the harm requires some kind of assignment of blame.[16] At a minimum, the case for a remedy is stronger when a person acts culpably rather than innocently, and so it is worthwhile to inquire whether the United States or Americans can be blamed for contributing to climate change. Indeed, the notion that Americans have acted in a blameworthy fashion by contributing excessively to climate change is an important theme in popular debates.[17]

Negligence in General

The weakest standard of culpability is negligence: if one negligently injures someone, one owes her a remedy. Economists define negligence as the failure to take cost-justified precautions. Lawyers appeal to community standards: a person is negligent if she did not take the level of care that a reasonable person would have in the same circumstances. Today, a scientific consensus holds that the planet is warming and that this warming trend is a result of human activity. But this consensus took a long time to form. In the modern era, the earliest work on global warming and greenhouse gases occurred in the 1950s, and the modern consensus is a product of the 1990s. Greenhouse gas emitting activities could not have become negligent, under existing legal standards, until a scientific consensus formed and it became widely known among the public—a fairly recent occurrence.[18]

Even today, it is not clear when and whether engaging in greenhouse gas emitting activities is properly characterized as negligent. The scientific consensus does not answer the critical question, for the purpose of determining negligence, of how much any particular activity actually contributes to climate change. Indeed, as we saw in chapter 1, a lively controversy exists about the overall costs and benefits of climate change in particular regions. Suppose, for example, that a large company in New York emits a large volume of greenhouse gases—is it negligent? It is easily imaginable that the costs of emissions abatement would be significant; it is also easily imaginable that the benefits of emissions abatement, in terms of diminished warming, would be close to zero. (Even very large emitters produce, in any particular period of time, little in the way of warming.) We all understand what it means to drive a car negligently so as to put other drivers and pedestrians at risk, but the claim that driving a (non-hybrid?) car carefully is in fact negligent because of its impact on global warming and the harm it causes to people living in India, is doubtful in light of the fact that the global warming cost of driving a car is trivial and the benefits, to the driver and others, may be significant. Heating a house, driving a car, running a freezer, taking an airplane—are all of these activities negligent? Even though the warming effects of the relevant emissions are infinitesimal?

It would be possible to respond that, in fact, negligence has been pervasive. Although the harm caused by each of these activities in isolation is small, the cost of precaution is also often low. For example, William Nordhaus calculates that, under certain assumptions, the optimal carbon tax as of 2010 would be about $34 per ton.[19] The calculation is based on the external cost of burning a ton of carbon as a consequence of greenhouse gas emissions. We calculate that this $34 per ton figure translates to about an extra ten cents per gallon of gas.[20] Using the economic theory of negligence as the failure to take cost-justified precautions, we could conclude that a person is negligent when she drives rather than walks if the benefit she obtains from driving is less than ten cents per gallon consumed. The argument could be extended to the choice of driving rather than using convenient forms of public transportation and to other activities as well.

Many people do seem to be reducing their emissions on the basis of an assessment of roughly this kind. Those concerned about climate change rarely believe that they should altogether stop engaging in activities that produce greenhouse gases (a difficult task!); instead, they think that they should cut back on activities that generate unreasonable emissions of greenhouse gases in light of whatever benefits they produce. Some people go further and purchase carbon offsets, but this type of activity seems, at present, supererogatory, whereas a case could be made today that a reasonable reduction of greenhouse gas emitting activities is morally required—that it represents an emerging community standard or norm.

Even if this is so, there is a problem with this argument, which is that the calculation given above assumes that everyone around the world, or at least hundreds of millions of people, are also cutting back on greenhouse gas producing activities. If many or most people fail to pay a carbon tax or (as we argue) fail to act as if they pay it by cutting back on less important activities that produce greenhouse gases, then the contribution of Americans who do this is quite small. And if this is the case, it cannot be considered negligent for Americans to fail to reduce their greenhouse gas emitting activities. Put differently, it is not negligent to fail to contribute to a public good if not enough others are doing similarly, so that the public good would not be created even if one did contribute.[21] This is a "moral collective action problem," and however it should be assessed in moral terms, the failure to act when other people are not acting, so that positive action would generate no benefit, does not seem to constitute negligence.

But our main point, for now, is that it is hard to argue that the stock of greenhouse gases in the atmosphere that can be attributed to the activities of Americans is the result of negligence, in either the moral or legal sense. Most of this stock was either produced before people understood the problem of climate change or was the result of activities, like heating one's house in the winter, that probably caused more good than harm, and in any event did not fall below community standards of care.

Negligent Government?

What about the U.S. government? Perhaps one could argue that U.S. climate change policy has been culpably negligent. The argument would be that, by failing to take precautions that would have cost the United States a lot but benefited the rest of the world much more, the U.S. government engaged in culpable behavior.

One problem with this argument is that, as we noted above, it is far from clear that the United States could have taken unilateral action that would have created benefits for the rest of the world greater than the cost to the United States. Unilateral reductions in greenhouse gas emissions would have little effect on overall climate change—not so far from zero even if aggressive and effective, and zero or very close to it if industry simply migrated to foreign countries. The Kyoto Protocol imposed no obligations on China, now the biggest emitter, and placed heavy burdens on the United States. So even if the United States had ratified that protocol, the effect would have been minimal. In this light, the claim that American policy has been negligent, under prevailing legal standards, is far-fetched.

It is also worth noting that, if an economic understanding of negligence is used, the United States does not emerge as a particularly culpable nation. On the economic understanding, people are negligent if their activities create costs that are greater than the benefits. If an economy delivers a unit of value in the form of goods and services, it produces benefits for consumers; if it does so while producing relatively few greenhouse gases, then those consumers may enjoy benefits greater than the costs to others. The greenhouse gas efficiency of an economy—in terms of minimizing costs on others for a given level of benefit—can be expressed as the ratio of emissions over GDP. As we saw in chapter 1, the most culpable nations using this measure are Zambia, Belize, and Liberia. The United States ranks 126th. So one might praise Americans who, over the years, have produced immense benefits for themselves at relatively little cost to the rest of the world. Or if one instead looks at emissions per capita, on the theory that emissions per capita rather than efficiency should be the proper basis for evaluating culpability, Belize, Guyana, and Luxembourg are the most culpable, with the United States ranked tenth. Is it plausible that a climate treaty should penalize Belize? Should people in Belize be given a disproportionate burden under a climate treaty because they or their government have not engineered a greenhouse gas–efficient economy? The principle of corrective justice cannot be applied selectively; if Belize should not be punished, then neither should the United States and Europe.

A more reasonable and serious criticism of American policy until very recently is that the U.S. government did not take seriously the risk of climate change, may have deliberately downplayed the risks when government officials knew better, and did not try to use its diplomatic power to advance climate treaty negotiations as much as it should have.[22] Maybe; but a reasonable alternative hypothesis is that the United States was just trying to exercise its bargaining power so that any eventual treaty would be more favorable to its interests than otherwise. It is farfetched to say that such common state behavior is negligent.

The Government versus the Public

Even if one could conclude that the U.S. government behaved negligently, it does not follow that the American people should be held responsible for their government's failures. The government itself does not have its own money to pay the remedy; it can only tax

Americans. To justify such a tax, one would need to conclude that Americans behaved culpably by electing or tolerating a government that failed to take actions that might have conferred benefits on the rest of the world of greater value than their costs.

There is a strong impulse to blame members of the public for the failures of their political system. In some cases, the impulse is warranted, but in others, the impulse should be resisted. The last example of such a policy was the war guilt clause of the Versailles Treaty, which held Germany formally responsible for World War I and required Germany to pay massive reparations to France and other countries. Germans resented this clause, and conventional wisdom holds that their resentment fed the rise of Nazism. After World War II, the strategy shifted; rather than holding "Germany" responsible for World War II, the allies sought to hold the individuals who formulated German policy responsible—these individuals were tried at Nuremberg and elsewhere, where defendants were given a chance to defend themselves. The shift from collective to individual responsibility was a major legacy of World War II, reflected today in the proliferation of international criminal tribunals that try individuals, not nations.

To be sure, no one is accusing the American government or its citizens of committing crimes; nor has the idea of a "climate guilt" clause surfaced so far. But the question remains whether Americans should be blamed, in corrective justice terms, for allowing their government to do so little about greenhouse gas emissions. It is one thing to blame individual Americans for excessive greenhouse gas emissions; it is quite another to blame Americans for the failure of their government to adopt strict greenhouse gas reduction policies. It is certainly plausible to think that voting for politicians who adopt bad policies, or failing to vote for politicians who adopt good policies, is not morally wrong except in extreme or unusual cases. Recall in this connection that even if Americans had demanded that their government act to reduce greenhouse gas emissions in the United States, the effect of unilateral reductions on climate change would be very small. And even if Americans had demanded that the United States lead the way on a climate treaty, it is unlikely that such a treaty would have produced much of a benefit, given the entrenched opposition of developing countries to greenhouse gas restrictions on their own industry, until very recently.

Rough Justice

However appealing at first sight, corrective justice intuitions turn out to be a poor fit with the climate change problem—where the dispute is between nations, and where an extremely long period of time must elapse before the activity in question generates a harm. This is not to deny that a corrective justice argument can be cobbled together and presented as the basis of a kind of rough justice in an imperfect world.[23] Even if not all people in developed nations are wrongdoers, and not all people in developing nations are victims, enough people in the first group are wrongdoers and enough people in the second group are victims, that transfers from developed nations to developing nations would do justice. The innocent payers should give up their rights so that the rights of so many others can be vindicated.

Perhaps the argument, while crude, is good enough to provide a factor in allocating the burdens of emissions reductions. Unfortunately, even that conclusion would rely on notions of collective responsibility that are not easy to defend. Most of the attractiveness of the corrective justice argument derives, we suspect, from suppressed redistributive and welfarist assumptions, or from collectivist habits of thinking that do not survive scrutiny.

Another argument along these lines is that because people take pride in the accomplishments of their nation, they should also take responsibility for its failures.[24] Americans who take pride in their country's contributions to prosperity and freedom should also take responsibility for its contributions to global warming. This argument, however, is especially weak. Many people are proud that they are attractive or intelligent, or can trace their ancestry to the Mayflower, or live in a city with a winning baseball team, but nothing about these psychological facts implies moral obligations of any sort. A person who is proud to be American, and in this way derives welfare from her association with other Americans who have accomplished great things, perhaps should be (and is) less proud than she would be if she were not also associated with Americans who have done bad things. She does not have any moral obligation, deriving from her patriotic pride, to set aright what other Americans have done wrong.

Corrective Justice: Taking Stock

In this chapter, like the previous one, we accepted for purpose of argument the underlying moral principles—that wealth should be redistributed to the poor, that victims of wrongdoing should have a remedy. But whereas in the last chapter we made an argument about means—a climate treaty is not a good way to redistribute wealth— here we question whether principles of corrective justice are relevant at all to the problem of climate change. Our conclusion is therefore stronger. While it is conceivable (but unlikely) that a climate treaty could turn out to be a good way to redistribute wealth, it is inconceivable that a climate treaty would properly address a problem of corrective justice—unless, of course, the questionable premises of the rough justice argument are accepted.

For many developing countries, the rich world's contribution to the climate problem is just one in a long series of grievances that include complaints about depredations committed by imperialist countries in their colonies, unfair trade relationships, interference with sovereignty, hard-hearted refusal to make life-saving drugs available or to give adequate foreign aid, and so forth. Some of these grievances have merit, others do not. Much depends on the particular details of the relationship between one state and another, so lumping states together into a wrongdoing rich world and a victimized poor world makes little sense. To the extent that some nations have legitimate complaints against others, those complaints can be addressed only through state-to-state negotiations that involve the original wrongdoers and victims and that culminate in reparations, assistance, or apologies that are appropriate to the original wrong. The notion that these complex problems could be globally addressed in a climate treaty is just not realistic and wishes away the serious disagreements that would first need to be resolved.

For the time being, it is sufficient to observe that the developed world acknowledges few of these complaints and has done little to address them. It would be tragic if climate negotiations broke down because developing nations believed that only a treaty that rich countries found unacceptable could address their accumulated grievances. The developing nations would suffer the most in the event of such a failure. In international relations, as in domestic politics, progress can be made on one front only if participants agree in the meantime to put aside their disputes on other fronts.

Chapter 6

Equality and the Case against Per Capita Permits

We have noted that many people believe that the problem of climate change should be handled by an international cap-and-trade system. Under this approach, participating nations, and perhaps the entire world, would create a "cap" on greenhouse gas emissions. Nations would be allocated specified emissions rights, which could be traded in return for cash. Though most economists favor a carbon tax, a system of this kind would probably be adequate and appears to be more politically feasible.

By itself, however, the proposal for a cap-and-trade system does not answer a crucial question: How should emissions rights be allocated? It is tempting to suggest that the status quo, across nations, provides the appropriate baseline. On one view, emissions should be frozen at existing levels, so that every nation has the right to its current level of emissions. On a more aggressive view, generally captured in the Kyoto Protocol, all or most signatory nations should have to reduce their emissions levels by a specified percentage, again taking the status quo as the foundation for reductions. The status quo might seem to have intuitive appeal, but it is also somewhat arbitrary and raises serious questions from the standpoint of equity. Why should climate change policy take existing national emissions, and to that extent existing national energy uses, as a given for policy purposes? Should a nation with 300 million people be given the same emissions rights as a nation with one billion people, or 40 million people, simply because the emissions of the three nations, at the current time, are roughly equal?

Raising these questions, many observers have strenuously urged that in an international agreement, emissions rights should be allocated by reference to population, not to existing emissions.[1] The intuition is that the atmosphere is a common resource of all humans and, therefore, every person on the planet should have the same right to use it; it should not matter whether people find themselves in a nation whose existing emissions rates are high. This intuition seems to reflect a commitment to a type of fairness—equality or equal division. Those concerned about the welfare of developing nations are especially interested in per capita allocations of emissions rights.[2] Why should a poor nation, with a large population, be required to stick close to its current emissions level, when wealthy nations with identical populations are permitted to emit far more? Why should existing distributions of wealth, insofar as they are reflected in current emissions, be taken as the foundation for climate change policy? More bluntly: Why should the United States be given emissions rights that are roughly the same as China's and dwarf those of India, both of which have much larger populations?

The significance of this controversy can hardly be exaggerated. Most notably, the per capita approach has been described as "the most politically prominent contender for any specific global formula for long-term allocations with increasing numbers of adherents in both developed and developing countries,"[3] including India, China, and as many as 130 other countries, and the European Union.[4] However, the United States has indicated discomfort with the per capita system, arguing that developing countries that are, or will soon be, industrial powers—including China, India, and Brazil—will have to accept significant mitigation obligations in a climate treaty. It is unlikely, we will argue, that a per capita system will satisfy the demands of the United States, one of the world's leading greenhouse gas emitters on a per capita basis. Meanwhile, the per capita approach remains the reigning political and ethical paradigm for the distribution of permits because it has been largely unquestioned.

Our goal in this chapter is to examine the per capita system, in terms of both principle and feasibility. Our examination suggests that its current prominence and popularity are undeserved. Advocates of per capita allocations are correct on one point: In principle, there is little to be said for basing emissions rights on existing emissions levels. The most plausible defense of this approach is pragmatic. Nations are unlikely to sign an international agreement if they will be significant net losers, and wealthy nations might lose a great deal from any approach that does not use existing emissions as the baseline for reductions. But this pragmatic point shows only that powerful nations might well veto approaches that are better in principle; it does not show that those nations are correct to do so. As a normative matter, an approach based on per capita emissions rights seems preferable to one based on existing emissions, and there are strong intuitive claims, rooted in welfarist and other arguments, on behalf of such an approach.

As we shall also see, however, a per capita approach runs into powerful objections. Some of these objections are similar to the arguments about distributive justice we made in chapter 4; others are new. As we argued in chapter 4, there is no reason on distributive justice grounds to tie a climate treaty to a new multilateral foreign aid program. Instead, redistribution should be done in the best way possible, which, given its complexities, is unlikely to be as part of a climate treaty. For the same reason, a per capita allocation of permits is not required on distributive grounds. In particular, permit allocations inevitably will go to governments, not people. There is no ethical reason linking payments to governments to a climate treaty; instead, such payments must be compared to other methods of meeting distributive obligations.

Even if we were to include distributive concerns within a climate treaty—say on the grounds that the creation of emission permits is a unique event that allows transfers that might not otherwise occur—a per capita allocation would be a poor method of doing so. Although there is a correlation between per capita emissions and wealth, some rich nations have low per capita emissions and some poor nations have very high per capita emissions. Many poor nations would be hurt, possibly severely. Moreover, any emissions reduction agreement will impose a disparate array of costs and benefits, varying greatly across nations; per capita emission rights do not take into account the variations in benefits from a climate treaty. From the standpoint of global redistribution—justified on grounds of either welfare or fairness—other approaches, more directly focused on the central goals, would be much better.

Many people support the per capita approach not on distributive justice grounds, but on the basis of a simple and plausible appeal to fairness.[5] The atmosphere's carbon-absorbing features are naturally thought of as a common resource. Perhaps a common resource should be divided among all the people in the world on the ground that all people enjoy a right to equal opportunity or to equal human dignity. Indeed, the same type of argument has been made about mineral resources discovered under the high seas: as no particular state "owns" these resources, they should be divided on a per capita basis.[6] And given the constraints of national sovereignty, the resources should be given to national governments on the basis of their states' share of the global population rather than divided up among individuals directly. We will argue that the analogy to common property is largely question-begging.

We shall also explore a series of welfare-related and pragmatic problems with the per capita approach, including its incentive effects with respect to future international agreements and population growth. A pervasive question involves feasibility. The problem of climate change cannot be successfully addressed without an international agreement that includes all or almost all of the major contributors. Per capita allocations would have the effect of redistributing hundreds of billions of dollars from wealthy nations, above all the United States, to developing nations. For this reason, insistence on per capita allocations would effectively doom any climate change agreement. We offer some brief remarks about the relationship between this pragmatic constraint and some of the underlying questions of principle.

Our conclusions are that on welfarist grounds the per capita approach is at most a crude second-best, and that it faces decisive objections from the standpoint of feasibility. Insistence on that approach would doom an international effort to reduce the risks associated with climate change. Further, the principle of equal division and the analogy to property rights have little appeal.

The Effects of a Per Capita Permit System

Before examining whether a per capita allocation of permits is required by ethical theories of climate change, we need to examine its effects. As we noted in chapter 1, per capita emissions go up with wealth. Therefore, as a general matter, a per capita allocation of permits would redistribute income from wealthy nations to poor nations. A simple back-of-the-envelope calculation shows that, as compared to a status quo allocation, the United States would lose hundreds of billions of dollars per year with a per capita allocation.[7] China right now has about the average per capita emissions, so it would not win or lose relative to a status quo allocation. India, which has very low per capita emissions, would be a big winner. A number of poor and very poor nations, such as Malaysia, Papua New Guinea, Zambia, Indonesia, and Brazil, have high per capita emissions and would face large losses.

Table 6.1 is a list of countries ranked by per capita emissions in 2000 for the six Kyoto gases plus land use change. (Unfortunately, we do not have more recent data that includes land use change. The 2000 data for China in particular are significantly out of date, and we include the 2005 data for China, excluding land use change, in brackets.) We include the top thirteen countries, some major emitters in the middle ranks, and a selection of countries with the lowest per capita emissions. The global average emissions are around 5.5 tons per person, so nations with emissions above that amount would have to purchase additional permits on the global market. A rough estimate of the net inflow or outflow for a country is the difference in per capita emissions and the global average (5.5 tons per person), multiplied by the permit price and its population.[8] For example, each person in the United States currently emits about 22.8 tons of carbon dioxide per year, so each person would, in effect, have to purchase permits for around 17 tons. If permits cost $50 and there are 300 million people in the United States, the total outflow from the United States to the rest of the world would be $255 billion per year. For Brazil, the outflow would be $75 billion per year.

A second important point is that anyone who favors a treaty that stabilizes greenhouse gas concentrations favors eventually moving

Table 6.1.

Per Capita Emissions in 2000, Six Kyoto Gases Plus Land Use Change

Rank	Country	Tons CO_2 Per Person	Population (thousands)
1	Belize	93.9	292
2	Qatar	53.5	796
3	Guyana	53.2	739
4	United Arab Emirates	38.4	4,104
5	Malaysia	36.6	25,653
6	Papua New Guinea	28.9	6,070
7	Kuwait	28.4	2,535
8	Australia	26.5	20,400
9	Antigua and Barbuda	25.3	83
10	Bahrain	25.2	725
11	Zambia	25.2	11,478
12	Canada	24.9	32,312
13	United States	22.8	296,507
...			
24	Indonesia	14.9	220,558
32	Russia	13.4	143,150
34	Brazil	13.3	186,831
47	United Kingdom	10.7	60,226
120	China	3.8 (5.5 in 2005)	1,304,500
...			
162	Pakistan	1.7	155,772
163	Kyrgyzstan	1.7	5,144
164	Mozambique	1.6	20,533
165	Yemen	1.6	21,096
166	Rwanda	1.6	9,234
167	Tajikistan	1.5	6,550
168	India	1.5	1,094,583
169	Burundi	1.5	7,859
170	Swaziland	1.5	1,131
171	Lesotho	1.4	1,981
172	Eritrea	1.4	4,527

toward roughly equal per capita emissions. The reason is that to stabilize concentrations, emissions will eventually have to get very low. To stabilize concentrations at a modest level, this will have to happen relatively quickly. Once emissions are low, differences in per capita emissions largely disappear; if everyone is close to zero, then everyone has about equal per capita emissions. The debate over per capita emissions, therefore, is a debate over how fast the world moves to equal per capita emissions. This does not make it unimportant, however—a thirty- or forty-year difference in timing can involve trillions of dollars of transfers. Nevertheless, it is important to recognize that at least among those who favor stabilizing greenhouse gas concentrations, the issue is about how quickly we make the transition to equal per capita emissions rather than a fundamental disagreement about whether we ever get there.

A number of commentators have supported a slow move toward equal per capita emissions on the theory that a slow transition reduces disruptions, calling this approach one of "contraction and convergence."[9] While these commentators purport to be in favor of equal per capita emissions allocations, they are not. Equal per capita emissions arise as an artifact of low global emissions. Because they favor a slow transition, in effect they reject the only important effect of equal per capita allocations.

There is a third and important effect of per capita allocations, which is that this approach does not allocate the net effects of a treaty on an equal basis. Although the costs would be allocated equally to all people, the benefits of a treaty are not equal. As we saw in chapter 1, the benefits of reducing emissions vary, depending on many factors, such as exposure to sea level rise or changes in weather patterns, dependence on agriculture, location of valuable mineral deposits, susceptibility to disease, and the likelihood of refugees from neighboring states. The net benefits under a per capita allocation, therefore, would not be equal.

For example, Thailand, Romania, and Jamaica all had roughly equal per capita emissions in 2000 and all are at the world average. If their emissions stay the same, they would have neither inflows nor outflows from a per capita permit allocation, at least in the initial years. These countries, however, may experience very different effects from climate change. Jamaica may be exposed to changes in hurricane intensity in the Atlantic. Thailand may face changes in agricultural patterns. Both are exposed to sea level change, while Romania is not. The net effects of a climate treaty with per capita allocations would be quite different for these countries. A similar comparison can be made between Papua New Guinea and Kuwait, which have about the same per capita emissions (both among the highest in the world), between Afghanistan and Vietnam (both very low emissions), or between any number of other nations. To the extent that the allocation of permits is driven by a claim that a climate treaty should produce equality, per capita allocation would not meet this goal.

The Per Capita Approach in Principle from a Welfarist Perspective

The Case for the Per Capita Approach

In discussions about climate treaties, defenders of the per capita approach argue that this approach is fairer than likely alternative approaches, such as the status quo approach. This argument is especially prominent in the developing world, where it is asked: Why should wealthy nations be given an entitlement to their existing emissions rights? This question seems to be one of fairness, to which we will turn in due course. But it can also be translated into a plausible welfarist argument, to the effect that the per capita approach is more likely to increase social welfare than any imaginable alternative. It makes sense to begin with the welfarist argument, which is in some ways more tractable, and which will illuminate the fairness questions as well.

Welfarists care about two things: maximizing the size of the pie and distributing it equally. The larger the pie, the more that is available for everyone to consume, and all else equal, welfare should rise with consumption.[10] At the same time, most welfarists believe that the welfare, or utility, that is obtained from an additional good is declining.[11] If you have zero apples, you are willing to pay a lot for one apple. If you have ten apples, you are willing to pay much less, or zero, for an eleventh. Thus, if the entire pie is given to one person, social welfare is not maximized. Ideally, the pie should be maximized, and then it should be divided into equal pieces, each of which is given to one member of society—but only assuming no disincentive effects, which might decrease the size of the pie. We can easily see that if disincentive effects are small, welfarists would advocate redistribution of resources from wealthy nations to poor nations, or at least from wealthy people in wealthy and poor nations to poor people in wealthy and poor nations.

There is an argument that the pattern of emissions allocations should not affect the size of the pie. The reason is that a cap-and-trade system gives individuals and governments incentives to minimize their emissions of greenhouse gases regardless of how permits are allocated. Regardless of whether a polluter owns a permit or needs to buy one, if emissions reductions are available at less than the permit price, the polluter would reduce emissions. If emissions reductions cost more than the permit price, the polluter would purchase a permit. Efficiency requires that marginal abatement costs be equal across all polluters; this will occur regardless of how permits are initially allocated. Said another way, optimal incentives will depend on the *quantity* of permits but not how they are distributed. As long as decision makers choose the right quantity, the size of the pie will be maximized. We can think of this as *ex post* efficiency, in the sense that it focuses on the efficiency effects of permit allocations once they are allocated.

Thus, the welfare effects of different permit allocation schemes depend on their distributional effects; other things being equal, distribution to those who are poor will increase welfare. The per capita approach might well seem to have attractive distributional effects and for that reason attractive welfare effects. To the extent that low emitting countries tend to be poorer, the per capita approach will help poor people, and because poor people have the highest marginal utility for a dollar, helping poor people will maximize global welfare. Certainly compared to the status quo approach, per capita allocations seem supportable on welfarist grounds; at first glance, they seem to be the right way to proceed. The examples of the United

States on the one hand, and India and many African countries on the other, are highly salient, because the former is rich and the latter are poor by comparison. To the extent that the per capita approach would require the United States to give hundreds of billions of dollars to India and many African countries, it might seem desirable on welfarist grounds.

Objections and Concerns

Distribution

The claim that permits should be allocated on a per capita basis for distributive reasons has the same set of problems that we discussed in chapter 4, which considered the general connection between distributive concerns and climate change. A per capita allocation of permits would be akin to an efficient climate treaty with side payments. As we discussed, there is no reason to connect a climate change treaty and policies designed to fulfill distributive obligations to the poor.

In fact, there are good reasons for not doing so; given how difficult and complex both problems are, the chances of getting both right diminish when we try to do both at the same time. This concern applies with particular force in the case of per capita permit allocations. Because nations would be the negotiating parties to any treaty, the governments would inevitably receive the permits, not the citizens. Even governments of well-functioning democracies may not fully represent all of their citizens, and most governments are far less representative than this. Nearly all poor states have a class of wealthy elites, and these wealthy elites usually control the government or have considerable influence over it. Given that the governments in these states already are unenthusiastic about redistributing wealth from the elites to the poor, it is questionable whether they will use the wealth generated by the permit scheme to help the poor. As a method of redistributing to the poor, a policy of permit allocations to the governments of low-emitting nations seems particularly poorly designed.

A second problem is that a per capita allocation of permits is not well targeted to the poor (even assuming that the governments of poor countries are representative of their citizens). Under a per capita system, permits would be distributed to both climate change winners and losers. As we noted above, some poor states will become far poorer as a result of climate change; others are less vulnerable. Some rich states will face serious adverse effects from climate change; others are less vulnerable. Some poor states, and some rich states, may even be net gainers from climate change. Ideally, permits should be distributed in light of these consequences, but the per capita approach fails to take them into account. If distribution is our concern, why should two equally populated poor nations receive the same number of permits from a program from which one gains a lot and another a little—or from which one gains a lot and another actually loses?

The key point is that the intuitive attractiveness of the per capita approach depends on seeing it in isolation from all of the effects of a climate treaty and from other global policies, including other policies with distributive effects. Once we take these factors into account, the per capita approach appears far less attractive, and on plausible assumptions, indefensible from the standpoint of the very accounts that at first sight justify it.

We agree that as a matter of actual practice, these defects are not necessarily fatal to the per capita approach. Everything depends on the alternatives. One might argue in response not that the per capita approach is ideal, but that it is superior to a system that is its most likely alternative—one that uses status quo energy consumption as the baseline and thus favors people living in wealthy and wasteful countries. Perhaps this response is correct. But it must acknowledge the underlying problem, which is that the per capita system is only indirectly connected to the underlying normative goal.

A welfarist should favor redistribution to the world's poor to the extent that doing so is feasible and does not excessively reduce the total size of the pie. But if one is a welfarist, there is no reason to think that the per capita approach to climate regulation is the right way to redistribute wealth and thus to increase global welfare. It would be much better to redistribute all resources than to redistribute shares of the atmosphere's capacity to absorb greenhouse gases; it would be much better to redistribute resources to poor people than to poor nations. From the welfarist perspective, a sensible redistributive policy would follow these general principles. If it is impossible—politically or technically—to redistribute all resources, then one needs to explain why it is possible—politically or technically—to adopt a per capita approach for a climate treaty.

Efficiency

Arguments in favor of per capita distribution have, so far, focused on what we have called ex post efficiency effects and neglected the possible ex ante effects of the distribution scheme. We discussed the ex post effects above, and showed that the ex post efficiency effects of the different schemes are identical (or nearly so). The same cannot be said for ex ante efficiency. From that standpoint, the effects are different, and the per capita approach has some significant drawbacks.

To understand the difference between ex post and ex ante efficiency, recall that any tax or cap-and-trade system that requires firms or individuals to internalize the social cost of their greenhouse gas emissions is efficient, in the sense that under these schemes firms and individuals will use energy only when the social benefits (including their own profits or consumption) are greater than the social costs (including the costs to the climate). We called this type of efficiency "ex post" because it addresses an existing problem, though, to be sure, one that will continue into the future.

The ex ante effect of a climate treaty refers to its effect on future programs, including those that have nothing to do with greenhouse gases. Any treaty will establish a precedent on which states will rely, at least in part, as they negotiate additional treaties in the future, treaties that will be needed to handle such global problems as terrorism, cross-border transmission of diseases, and nation-building efforts in failed states. For example, if the per capita approach is used for a climate treaty, then it will suggest itself as a basis for allocating the costs of a terrorism treaty.

Similar assumptions are routinely made about domestic programs. For example, the U.S. government could alleviate poverty by announcing one day that it will take most of the wealth of rich

Americans and give it to poor Americans. Such a program is not inefficient in the ex post sense: given that the rich have already accumulated their wealth, they cannot retroactively be deterred from working hard. The program will have prospective effect, however. Even if announced as a one-time event, people will assume that if the government implements such a program today, it might do so again tomorrow. This assumption will influence people's ex ante behavior, reducing their incentive to work and save.[12]

Suppose, then, that a climate treaty based on the per capita approach established a precedent. How might such a precedent influence behavior, compared to the baseline status quo approach? It would create two perverse incentives.

First, the per capita principle would establish that the most highly populated states would obtain the greatest benefits from international cooperation. Governments would be rewarded for pursuing fertility policies that maximize the size of the population.

To see why, consider a state with population X and another state with population 2X. Suppose that a future treaty would limit the spread of infectious diseases, creating benefits of Y. The states would need to negotiate a division of the surplus. With the per capita principle in place, the state with the larger population would be able to claim a larger portion of the surplus. Note that the problem is not that the per capita approach would necessarily encourage states to increase their populations in order to obtain more permits; that incentive can be avoided with a stipulation that the number of permits is fixed on the basis of the population at the time of treaty ratification or shortly before it. The problem here is one of creating socially harmful incentives to improve bargaining position for subsequent treaties that need have no relation with climate policy.

From a redistributive perspective, this result might seem fair (unless the people in the larger state are richer), but in terms of prospective incentives, states now have one more reason to grow and to avoid shrinking. This incentive is especially perverse from the perspective of climate change, because more people will consume more of the earth's resources (though, conceivably, more efficiently). On the other side, the climate treaty, to the extent that it fixes the initial number of permits, could restrain population growth. And it is true that given the relatively limited amount of international cooperation, one might doubt that the incentive to expand population is particularly strong. To evaluate the extent of the problem, we need to know the magnitude and not merely the direction of the incentive effect. Still, it is a cost of the per capita system that should be kept in mind.

Second, to the extent that the per capita approach is used in the first place only because it favors poor countries, and hence the real principle is that poverty, not population, entitles countries to better treatment in treaty negotiations, governments that adopt sensible policies that promote economic growth would be penalized.

This incentive is also perverse. Most states get rich because they have good institutions, not because they are lucky enough to have natural resources.[13] Citizens invest in creating and maintaining good institutions because good institutions deliver wealth and other benefits. A redistributive principle such as the per capita rule implicitly punishes states that do well, while rewarding states that do poorly.

The goal of development aid over the past decades was precisely the opposite: to give governments of developing countries an incentive to adopt sound economic policies that promote growth. Because of fears that foreign aid would provide incentives not to grow, donors made concerted efforts to condition aid on the adoption of sensible growth policies.[14] The per capita principle—indeed, any redistributive principle—is at war with the lessons of development policy and would weaken the pro-growth incentives that are currently given to developing states.

What system, then, is optimal for ex ante efficiency? The ideal principle would give states an incentive to identify global problems in advance and negotiate treaties to solve them, and otherwise not affect their incentives to control their populations, invest in institutions, and so forth. Such a principle would be, at a minimum, a form of International Paretianism, so that states believe that they will not be made worse off by a legal solution, a belief that would discourage states from entering treaty negotiations.

But treaties that solve problems generate surpluses beyond the amount necessary to make states indifferent between entering and not entering a treaty. What should be done with the surplus? It is tempting to think that one can distribute the surplus without affecting incentives ex ante, but this is highly implausible. (If one can, then one would probably want to distribute the surplus to the poorest countries rather than on a per capita basis, which, as we have been arguing, is morally arbitrary.)

From an efficiency perspective, the best use of the surplus would be to reward the states that had taken steps in advance of the treaty to abate greenhouse gases.[15] These states would probably be the European states that accepted binding reductions under the Kyoto Protocol, though there are complexities here, since not all European states accepted meaningful reductions and others were simply taking advantage of independent technological and demographic changes in their country.[16]

The larger point is that such a distribution would establish a precedent to the effect that when a global problem exists, states that respond quickly and in advance of a treaty will not be penalized. With this principle in place, states would be more likely to act quickly and to negotiate a treaty regime rather than drag their feet. For example, if states ever need to enter a new treaty that regulates cybercrime, they will know that first movers that have implemented controls that reduce dangers to other states will not be penalized. Instead, the treaty will ensure that these states will be rewarded in some way.

Ex ante efficiency does not favor the per capita approach, but it also does not favor the status quo approach. Under the status quo approach, states that have acted least aggressively to reduce emissions do better, all else equal, than states that have acted more aggressively. We will address this problem in chapter 8, where we offer a modified version of the status quo approach that establishes a precedent favoring states that move first to address a global problem.

The Per Capita Approach from the Perspective of Fairness

Ideas about fairness are playing a significant role in debates over the proper approach to climate change. Fairness can be specified in multiple different ways. We present three specifications here in an effort to see whether the per capita approach can be defended on fairness grounds.

As we noted in chapter 5, many people reject the idea that questions of global justice should be approached in welfarist terms. In their view, the goal is not to promote aggregate social welfare; it is instead to do what fairness requires, which can be specified using the "veil of ignorance" method. Consider a commonsense specification of this claim, adapted to the climate change problem. Some nations are much richer than others, in a way that violates the requirements of justice. Perversely, the status quo approach creates a kind of entitlement to the continuation of practices that violate those requirements. No such entitlement can be defended. Even if corrective justice does not require high emitting states to compensate those nations that are at special risk, a climate change agreement would be unacceptably unfair if it makes it more difficult for poor nations to develop—especially because development is designed to remove their citizens from difficult conditions and to achieve something closer to the threshold or to equality with wealthy nations. A per capita approach is the most fair, because it allows every citizen to count for no less and no more than one, in a way that respects the moral irrelevance of national boundaries.

We do not intend to challenge these general points about fairness here. Our basic claim is that if they are taken as a defense of the per capita approach, they run into serious difficulties. The reason is that the central objections to the welfarist argument rematerialize when fairness, understood in the stated way, is our guide. To the extent that some of the most populous states are wealthy, the per capita approach is not fair at all; to that extent, it has some of the same vices as the status quo approach. Per capita allocations also have the disadvantage of giving large numbers of permits to highly populated nations that have relatively little to lose from climate change. And it remains true that permits are allocated to the governments of poor states, not to the citizens of poor states, and allocations to such governments may not help those who are most in need. If fairness requires redistribution across national boundaries, the status quo approach runs into significant trouble, and the per capita approach is better; but those interested in global redistribution would hardly choose that approach among a menu of possibilities.

Equality and the Atmosphere as Common Property

There is another type of fairness argument, to the effect that the atmosphere, with its beneficial carbon-absorbing characteristics, is common property, belonging to everyone in the world.[17] A climate treaty closes a commons, converting it to private property. It is only fair to distribute the parcels of property to the former users of the commons, namely, everyone in the world, on a per capita basis. One might draw an analogy to minerals discovered in the sea bed under the high seas, which are outside the sovereignty of any country. The Convention on the Law of the Sea provides that revenues from exploitation of these minerals should be distributed "equitably," although that term is not defined.[18]

The analogy to property is superficially appealing but, on reflection, it turns out that it does little but muddy the waters by creating an unnecessary set of abstractions between normative goals and policy outcomes.

In law, a commons refers to a resource to which numerous people have a legal right to access. If the government owns the commons and seeks to convert it to private property, it merely blocks people from using it. If the people who use the commons have a legal right to use it, then the government—at least, in most modern legal systems—would have to compensate them for giving up their rights. Sometimes, the legal rights might be derived from customary use; in such a case, the government would probably have to compensate people deprived of their customary rights. If the closure of the commons generates revenue for the government, the government would have no legal obligation to funnel it to all citizens on an equal basis; nor does the government have an obligation to distribute the resulting pieces of private property to everyone on an equitable basis.

So much for law; what of ethics? Of course, one could argue that the government has an ethical obligation to distribute the costs and benefits from enclosure in a fair way. These ethical obligations are not specific to enclosure itself; they are the same ethical obligations that a government must follow when it does all the things that governments do—tax people, spend revenues, and so forth. The history of enclosure provides little ethical illumination. In British history, enclosure was controversial mainly because those who sought enclosure (usually, private individuals with land subject to customary rights enjoyed by locals) tried to avoid compensating people for the loss of their historical rights. Supporters of enclosure believed that it led to the more efficient use of resources; critics believed that it harmed those who held customary rights, often the poor.[19]

Currently, everyone has access to the atmospheric commons. Does it follow that when the commons is closed, everyone should have equal access to it? Certainly, the appeal to law or history does not help answer this question. When governments close commons, they do not—as far as we know—distribute shares of it to citizens on a per capita basis. As noted above, the beneficiaries could just as well be customary users of the commons. If the commons is a pasture, the normal instinct will be to compensate people who have used the pasture for their animals. The reason is that those people have made investments in reliance on continued access to the commons, and it would be unfair to deprive them of investments based on reasonable expectations about the continuation of existing property rights.

If that pattern were followed for the atmosphere, then the closing of the atmospheric commons would generate entitlements to its customary users—namely, the people who have historically emitted the most carbon. The principle would be the same as in the pasture case: those people who have arranged their lives around patterns of carbon-intensive production—for example, by moving to spread-out cities where automobiles are the cheapest form of transportation—are most vulnerable to a radical change in property rights, and they should be compensated for the lost investment that they made in the reasonable expectation that rights would continue as in the past.[20]

The response is that the ethical claims of the poor should trump the claims of traditional users. We sympathize with this argument but do not see what it has to do with ethical claims based on the nature of common property. Indeed, the argument is that property should be redistributed to the poor *despite* claims derived from the enclosure of a commons, not *because of* them. The better approach is to abandon the analogy to common property and address the ethical issues directly.

The commons argument also misleadingly draws one's attention to features of the environment that do not "belong" to any particular state. The atmosphere is a commons because no state has a right to the atmosphere; by contrast, forests are not commons, in an international relations sense, because they belong to the particular states on whose territory they are located. The commons argument suggests that the two should be treated differently: states with forests have no obligation to share the carbon-absorbing value of those forests with other states. From an ethical standpoint, this distinction cannot be sustained. People should not gain or lose because of the contingency of the location of their birth, and so there is no ethical reason for giving states control over resources that happen to be located on their territory.

The implications of such an argument are dramatic. States that enjoy valuable natural resources—oil reserves, like Saudi Arabia, or diamond deposits, like Botswana—would have no right to the revenues from those resources. They have an ethical obligation to share those revenues with everyone in the world on a per capita basis. From an ethical standpoint, all people in all states would need to pool the value of their natural resources, along with the value of common natural resources such as the atmosphere, and distribute that value on a per capita basis to everyone in the world.

All of this suggests that the reasons that states enjoy the value of their own natural resources lie elsewhere. Perhaps if states had to share these revenues, they would not have sufficient incentives to exploit resources efficiently, and the world as a whole would be worse off. Or perhaps states believe that the beneficiaries of such a redistribution would squander their transfers rather than use them to help the poor. Or perhaps states simply refuse to engage in the type of massive redistribution that such an ethical theory implies because the people living in those states are selfish.

If these reasons explain why states do or should enjoy the value of "their" resources, then people making the commons argument need to explain why pragmatic reasons should not apply there as well. If states that would receive an abundance of permits under the per capita approach are unlikely to use these transfers wisely (just as they have failed to use foreign aid wisely in the past), then there would be no ethical obligation to adopt this approach. Or if states for self-ish reasons are not going to agree to such massive transfers, then we need to think about how to solve the climate problem in a way that even selfish states would agree to. If relative power puts limits on realistic ethical claims about how states should share their resources, then it also puts limits on realistic ethical claims about how the at-mospheric commons should be distributed. The appeal to common property simply does not help with this difficult issue.

Our limited point is that the analogy to common property does not provide an independent argument for per capita distribution of emission permits. It simply distracts from the ethical and pragmatic issues that a climate treaty must address.

Frugality, Profligacy, and Related Issues

Suppose that we understand the idea of fairness not in redistributive terms, but as a requirement that similar people be treated similarly. As we saw above, the per capita system is not attentive to the differential distributional effects of climate change and abatement costs, but in effect gives every person the same asset. From one perspective, the main objection to this feature of the per capita system is that it means that wealth does not necessarily go to the poor. But holding wealth constant, it might seem unfair that frugal individuals who have pro-duced few greenhouse gas emissions receive the same payout as profli-gates who have produced many. And it might seem unfair that people who are most hurt by climate change receive the same payout as those who are least hurt (or even benefited) by climate change. Finally, we might think people who are most hurt by the abatement efforts mandated by the climate treaty should receive some kind of compensation. Consider, for example, low-income workers who commute to work and must pay higher bus fares or fuel prices. One might argue that fairness requires that these people receive permits, so that they do not bear a disproportionate cost of the treaty regime.

These considerations obviously do not point in the direction of the per capita approach. But it also seems likely that any attempt to account for all such considerations would quickly get bogged down in wrangling over the details. One might argue, then, that the per capita approach might make sense as a second-best standard or kind of rough justice, along the lines of what we discussed in chapter 5. We turn to that argument next.

The Per Capita Approach as a Second-Best Standard

We have seen that in principle, significant global redistribution is plausibly justified by considerations of both welfare and fairness. But in practice, such redistribution is not occurring; for example, there is no evidence that the United States wants to transfer hundreds of billions of dollars to poor people in poor countries (existing foreign aid, most of it tied to particular types of reciprocal action, is much less). In these circumstances, defenders of per capita allocations might argue that their approach has three virtues. First, the per capita approach might be feasible even if a preferred form of redistribution is not. Second, such an approach might provide the basis for a kind of incompletely theorized agreement among those who have different moral commitments, or who are unsure about the appropriate moral commitments in the international domain. Third, per capita allocation might, because of its simplicity and attractiveness, provide a plausible focal point for political action—a basis for an international agreement to which many nations could subscribe, even if it would be fanciful to suggest that wealthy nations might sign an international agreement in which they agree to transfers hundreds of billions of dollars to poor nations. This is another type of rough justice argument.

Suppose nations acknowledge that certain moral principles guide international relations, or should, but that they disagree about what those moral principles are. If one believes the rhetoric of governments, one can identify a set of standard moral arguments. Among developing nations, some argue that the rich world has obligations to the poor arising from the history of colonial exploitation.[21] Others argue that rich nations have obligations arising from particular policies that they have adopted in the recent past and that continue in the present—unfair tariffs that discriminate against agriculture, for example, or immigration rules that drain away poor nations' educated elite.[22] Still others argue simply that resources that exist outside the sovereign territory of each state should be shared.[23] Some rich nations are willing to acknowledge that they have an ethical obligation to provide aid to the very poorest people; others say that they have an obligation to cooperate with poor nations or not to interfere with them but not necessarily to give them aid.[24]

These different moral arguments have different implications. Even among the poor nations, whose views seem consistent at first sight, one can detect radically different implications of the different arguments. If one focuses on colonial exploitation, then the major beneficiaries should be former colonies (including rich states like Taiwan) and the major payers should be former empires (including Great Britain, Russia, and Portugal but not so much the United States). The idea of colonial exploitation suggests that former colonies should direct their claims at their former masters, not to the rich world as a whole. India's extra permits, for example, should come out of Great Britain's pocket. Similarly, if tariff policy is the source of complaints, one would need to determine which tariff policies were supported by whom, and which countries they harmed—and this is highly complex and controversial. And if tariff policies that have adverse effects on other nations (and what tariff policies do not have such effects?) should count, so should all other policies that have given rise to legitimate grievances. One would thus need to keep in mind the particular grievances that some poor countries have against other poor countries (India and Pakistan, Rwanda and Burundi) and allocate permits accordingly.

It would seem that even if the rich nations owe extra permits to poor nations, within the class of poor nations, permits would have to be distributed unequally to account for current and past injustices. Generous treatment, such as the rich nations' contributions to the victims of the recent South Asia tsunami, would need to be subtracted, lest rich nations hoard their generous impulses as offsets to permit regimes. And all of this would need to be done in a manner that respected the views of those who care about redistribution on grounds solely of distributive justice or welfare maximization.

Within countries, moral disagreement of this type does not necessarily preclude policy, even on issues that divide people sharply along moral lines. Typically, the policy that emerges reflects an incompletely theorized agreement.[25] People with different moral views agree on a policy that is consistent with their different interests and different moral views, while bracketing their remaining conflicts or putting them off until a later time. For example, in the United States some people support affirmative action as a way to overcome past injustices, while others defend it as a forward-looking policy for promoting certain social goals, such as stability. The moral views have different implications for how affirmative action should be designed and how long it will last, but those holding these different views can sometimes agree on enough to put their weight behind a program that furthers some of their goals but not others. Similarly, one might argue that the per capita approach could reflect an incompletely theorized agreement among nations and individuals with different but overlapping moral views about what nations owe each other.

This argument also is weak. None of the moral views described above would support the claim that greenhouse gas permits should be distributed according to population size, with the possible exception of the view that commons should be shared. But even that view does not clearly distinguish between per-nation sharing and per capita sharing. If there is a common thread among these theories, it is the view that richer nations have an ethical obligation to aid or cooperate with poor nations. But as we have seen, poor nations and populous nations are not the same.

Feasibility Issues

Thus far our focus has been on issues of principle. There is also a question of feasibility. The appealing features of the per capita approach—its simplicity, its apparently clear appeal to intuitions about fairness—come at a high price. Scientific and economic models indicate that, most likely, substantial cuts in greenhouse gas emissions will produce global benefits in excess of global costs. But it is obviously difficult to obtain agreement on emissions reductions if some nations are likely to benefit far more than others from such an agreement. If a specified level of reductions will give significant benefits to India, but more modest benefits to the United States and Russia, the latter nations might well be reluctant to accept that level of reductions, and might demand some kind of compensation. Even more troublesome, restrictions on greenhouse gas emissions will probably be most costly for large emitting nations, including the United States. Large emitters, facing significant costs from emissions reductions requirements, therefore will be unlikely to join a treaty unless the treaty uses their status quo emissions as the baseline from which to determine cuts. As a first approximation, nations care about the welfare of their own citizens, and the welfare of citizens in other places is not a primary consideration and may not matter greatly.[26] A workable climate treaty will have to be one that serves the interests of the United States and other major industrial nations, including developing nations such as China and Brazil. As a practical matter, nations that are already the biggest greenhouse gas emitters will therefore not join a treaty that requires them to reduce their emissions to the level of very poor nations; nor would they enter a treaty that requires them to pay a lot of money for permits distributed to the poor nations.

Consider a few numbers in this regard. In 2006, the United States distributed almost $24 billion in foreign aid (a third of which was to Iraq).[27] The politically unacceptable Kyoto Protocol would have cost the United States a substantially larger sum each year over the indefinite future—the equivalent of perhaps tens of billions of dollars per year. The per capita approach (as compared to the status quo approach) would cost the United States far more than that—possibly more than $200 billion per year for the indefinite future.

We acknowledge the tension between feasibility arguments and ethical arguments. We try to resolve this tension by making the following two assumptions. First, only a treaty that satisfies International Paretianism—that is, that advances the interests of all states relative to the status quo—is feasible. Second, among the many treaties that satisfy International Paretianism, ethical principles will have some sway. Any feasible treaty creates a "surplus"—equal to the gains from mitigating climate change minus the costs of abatement—and this surplus can be distributed according to ethical principles. The distribution can be built into the abatement obligations themselves, or could be effected through side payments. We will say more about this approach in chapter 8; for now it is sufficient to observe that the per capita approach falls so far short of International Paretianism that it has no chance to succeed. To insist on the per capita approach, then, is most likely to subvert the best chance for a climate treaty and hence to render the climate change problem intractable—a special difficulty for poor nations that are particularly vulnerable to that problem.[28]

Fairness and the Per Capita Approach: Taking Stock

We have made two distinct points in this chapter. First, equality—in the sense of dividing equally, on a per capita basis, the "surplus" from a climate treaty—should be, at best, a weak consideration in the design of a climate treaty. It would be much better to design a treaty on the basis of fundamental normative principles, while taking into account feasibility constraints that arise from the state system. Second, the per capita approach, in particular, does not deserve as much prominence as it receives. It falls short on the basis of a number of normative criteria, including equality and distributive justice, and it is infeasible as well.

Chapter 7

Future Generations
The Debate over Discounting

Any analysis of the ethical issues associated with climate change must come to terms with the fact that the benefits of emissions reductions will be enjoyed in the future rather than the present. If the world cuts emissions immediately, the beneficiaries of its action will be people living decades from now, not people living today. By contrast, the costs of emissions reductions will be paid mostly by current generations. How should policymakers and analysts deal with future benefits and present costs? Among economists, the standard answer is that future effects should be "discounted." A dollar today is worth more than a dollar in a year. Economists ask: Shouldn't the future benefits of reduced climate change also be discounted?

The debate over this question might seem to be a technical, mathematical issue, but it turns out to be one of the central ethical dilemmas in evaluating climate change. Seemingly small changes in the discount rate can lead to very large changes in estimates of the costs of climate change and the benefits of abatement. With a high discount rate, the argument for immediate, aggressive action to reduce climate change seems weak. With a near-zero discount rate, that argument seems extremely strong.

Defenders of discounting argue that discounting at the market rate of return is necessary to ensure consistent comparisons of resources spent in different time periods. Critics of discounting begin with a principle of intergenerational neutrality. They insist that people in the current generation should not be treated as more valuable than people in the next generation; they object that discounting ensures that future generations will receive less attention, and perhaps far less attention, than those now living. If we discount money at (say) a rate of 5 percent, people living in 2150 would seem to be worth merely a small fraction of people living today.

Discounting has generated a large literature with strong views on various sides of the issue;[1] one of our main goals is to explain those views and to untangle some of the ethical issues. As we shall see, the issue of discounting turns out to be closely analogous to that raised by the distribution of costs and benefits across countries, except with discounting, the distribution is across time instead of across space. We will argue that the same analysis applies to the two cases. In particular, we argued in chapter 4 that we should not think of climate change abatement as the best means of redistribution if more direct methods of redistribution are available. The same argument applies across time. We should ensure that our decision on how much to invest in climate change abatement is efficient and examine direct methods of redistribution across generations, such as increasing or reducing the overall set of investments for the future more broadly. If we make a decision to invest in a particular project, such as climate change abatement, using a lower discount rate, we miss the opportunity to make similar investments that have a higher rate of return.

Along with the critics of discounting, we accept the principle of intergenerational neutrality. If decisions based on cost-benefit analysis with discounting impose serious harms on members of future generations, then there is indeed a serious ethical problem. As we shall show, it is possible that discounting will have this unfortunate result. The solution is not, however, to refuse to discount at the market rate of return when choosing a particular project. It is to ensure that the overall set of policies give greater weight to the interests of future generations. As we shall also show, a refusal to discount can, under reasonable assumptions, harm rather than help members of future generations by depriving them of the benefits of current investments. Our conclusion, then, is that discounting at the market rate of return is appropriate and ethically justified; when cost-benefit analysis with discounting produces indefensible results, the response should be not to refuse to discount at the market rate, but to invest more for the future, choosing projects with the highest available rate of return.

The market rate of return is the rate of return on capital investments broadly construed; if you were to invest a dollar in the overall market today, it is the rate of return on that dollar. A significant problem with computing this rate of return is that there is significant uncertainty about what this rate will be in the distant future; the effects of climate change, for example, may significantly change the available rates of return, but we cannot know by how much. When we calculate the expected market rate of return under these circumstances, it turns out as a matter of mathematics that we get a very low number (implying that we should give relatively substantial weight to future payoffs). For reasons we will illustrate, low rates of return dominate the calculation. The result is that even those favoring discounting at the market rate should support discounting at a low rate.

Why Discount?

The basic problem addressed by discounting is that the costs and benefits of spending resources to reduce the effects of climate change come at different times. To prevent significant harms, we have to begin spending sizable resources in the near future. The benefits of these expenditures, however, will be enjoyed over the next several hundred years in the form of reduced effects from climate change. Different people will pay the costs and receive the benefits, just as in the cross-country case.

For short-term projects where the costs and benefits occur in different but relatively proximate periods, it is standard to discount the costs and benefits to a single period. Suppose that you have $100 and are presented with a $100 investment that will produce $110 in two years. You can choose to spend the $100 today or make the investment and spend $110 in two years. You should not simply choose $110 on the grounds that $110 is larger than $100; that would be ridiculous. Instead, you should use discounting to compare the two choices.

To see why, suppose that you have an alternative choice: putting the money in a bank account, which will pay you interest at 6 percent. If you put the $100 in the bank instead of making the investment, you would have $112.12 after two years. Therefore, you should not make the investment. Doing so would leave you worse off than the alternative. The equivalent procedure for making this comparison is to discount the investment returns by the interest rate. If the discounted flow is more than $100, the investment makes sense and if not, it does not. To discount, simply divide each cash flow by $1/(1+r)^n$ where r is the interest rate and n is the number of years. Here, the present value of $110 when the interest rate is 6 percent is about $98 ($110/1.06^2 = 97.90), which is less than $100.

Discounting is best seen as a way of taking opportunity costs into account. The opportunity cost of making the investment is the alternative return available by putting the money in the bank. With our numbers, the opportunity cost is higher than the return on the investment. If the bank paid interest at 4 percent, the opportunity cost would be lower—and, in this case, the investment would make sense.

A key to this analysis is that you had alternative means of shifting resources across time, in our example a bank account. Given the bank account, you could reasonably choose $100 today or $112.12 in two years, but in no event should you choose $110 in two years. If you were Robinson Crusoe and had no other choice of investments—no way of shifting resources across time periods by borrowing or by saving—the opportunity cost of the investment would be zero and you would simply have to decide whether you preferred $110 in two years to $100 today. You might still prefer $100 today because you have to eat, but you would no longer look to the opportunity cost to decide. As we will see, an important implicit disagreement in debates about climate change is whether the world is like Robinson Crusoe with respect to climate change—whether opportunity costs should be ignored or taken into account.

Future Generations and the Arc of Time

A key question is whether this same logic can be applied over very long time periods (perhaps 200 years or more) and when the costs and benefits are spread across generations rather than a single individual. Discounting is no longer about an individual spreading consumption over a lifetime. Instead, it is about comparing the welfare of different individuals. Does the same logic apply?

It is clear that discounting can have profound effects on policy choices that have long time horizons. Suppose that as a result of climate change, we are facing the loss of $1 trillion in one hundred years. If the discount rate is 7 percent, we would be willing to spend only a little over $1 billion—one one-thousandth of the damages— to prevent that harm! If the time horizon were two hundred years—a time well within climate change policy considerations—and we use a 7 percent discount rate, we would be willing to spend only about $1.3 million, 0.00013 percent of the future cost, to prevent it. With these numbers, it is easy to construct examples where it is not desirable to spend a small amount of money today to save some valuable asset for a large number of people in the future. On ethical grounds, many people are skeptical about the conclusions seemingly required by discounting the future benefits of reductions in climate change.

We have noted that discount rates are central to major disagreements over climate change policy. For example, two of the most prominent recent studies of climate change had widely different recommendations. The Stern Review, a study commissioned by the UK government and headed by Nicholas Stern, recommended very high taxes on emissions and dramatic and fast reductions. William Nordhaus, a prominent environmental economist, recommended modest taxes and a slow phase-in of reductions. Almost the entire difference between these influential recommendations is driven by the discount rate assumptions used by the authors. Stern discounted future costs and benefits at 1.4 percent, while Nordhaus discounted them at 5.5 percent.[2] A 1.4 percent discount rate would value a cost in one hundred years almost fifty-three times as highly as would a 5.5 percent discount rate. If the harm occurs in two hundred years, the Stern approach would value it almost 2,800 times as much as the Nordhaus approach. If most of the costs of climate change occur in the distant future, the discounting assumption would lead Stern to see climate change as a far larger problem than Nordhaus does. To an approximation, discount rates are all that separated the authors.[3]

Analysts have long recognized that discount rates are among the central parameters in evaluating the effects of climate change and that the decision about the appropriate response depends on resolving the debate. The IPCC, for example, estimates that discount rates are the second most important factor in evaluating the effects of climate change. The IPCC rates the effect of climate sensitivity (the average global temperature change due to a doubling of the concentration of CO_2 in the atmosphere) as the most important factor. If this factor were scaled at 100, discounting would have a value of 66, while estimates of the valuation of the economic impact from a 2.5 degree increase in temperature is valued at 32.[4]

The debate on discounting goes back to long before climate change was an issue. Just looking at prominent economists, we can see widely different viewpoints. In 1928, Frank Ramsey famously argued that "it is assumed that we do not discount later enjoyments in comparison with earlier ones, a practice which is ethically indefensible and arises merely from the weakness of the imagination."[5] Roy Harrod argued that discounting is "a polite expression for rapacity and the conquest of reason by passion."[6] Tjalling Koopmans, a Nobel Prize winner, argued, however, that a failure to discount effectively means that the current generation must starve itself to benefit the future. Suppose that ethically, we must act to maximize total welfare and that a dollar invested grows at a positive rate for the indefinite future. If we do not discount, the future gains from the investment dollar will also be worth infinitely more than the present loss. Therefore, not discounting means that we must save every dollar, which is an indefensible conclusion.[7] Numerous other authors have studied this issue over the years.[8]

Ethicists and Positivists

There are two major positions with respect to the proper discount rate. We will describe these as the ethicists' and the positivists' approaches. The ethicists attempt to reason from first principles about what the discount rate should be. The positivists attempt to observe what the market-determined discount rate actually is. If the market rate does not coincide with what the ethicists think it should be, the two positions will conflict. We begin by describing the approach of the positivists and then turn to that of the ethicists. In the end, of course, the positivists' approach is worth nothing unless it can be defended on ethical grounds.

We will suggest that the two approaches focus on different issues. The ethicists' approach should not be used to choose projects with a lower-than-market rate of return, and the positivists' approach should not be mistaken for a moral judgment about our obligations to the future. We suspect that the intuitions behind the ethicists' arguments turn on a conflation of discounting welfare with discounting money. The ethicists are correct to say that the welfare of a person born in 2050 does not matter less than the welfare of a person born in 2010. But as we shall see, they are wrong to suggest that this claim makes it unacceptable to discount money.

The Positivist Position

The positivists approach the issue as a simple problem of opportunity costs, even for the long term. Suppose that we were going to invest $100 billion to reduce carbon emissions, producing a benefit in one hundred years of $400 billion. This represents a rate of return of 1.4 percent, the discount rate used in the Stern Review. The positivists reason that if the market rate of return over that time period is 5.5 percent (the Nordhaus rate), the same $100 billion could be invested to produce over $21 *trillion* in one hundred years, almost fifty-three times as much. Equivalently, we could give the future $400 billion by investing about $2 billion at the market rate, keeping the remaining $98 million to spend on riotous living now. It does not make any sense, argue the positivists, to invest the $100 billion to reduce the effects of climate change under this hypothetical set of facts. To do so would be to throw away vast resources: either $98 billion today or more than $20 trillion in one hundred years.

The conclusion, following this logic, is that it is not sensible to invest in any project unless it has a return at least equal to the return available elsewhere. The problem is exactly parallel to that of the individual who compared the return on the investment to the return available elsewhere (the bank). The long time period does not change the method of analysis; it only makes the issue more important. We should, therefore, discount projects at the otherwise available return—the market rate of return. Only projects that pass discounted cost-benefit analysis should be undertaken. Any other choice throws away resources, which is not good for future generations. In our example, instead of investing the $100 billion in the project that had a 1.4 percent return, we could invest, say, $5 billion in the market and give the future generation about $1 trillion. Everyone would be better off: the current generation would have $95 billion more than otherwise, and the future would have $600 billion more than otherwise.

Note that the positivists are discounting money; they are not necessarily treating people living in the future as worth less than people living today. Discounting money and attributing equal moral worth to people of different generations are fully compatible. To see why, suppose that a statistical life today has a value of $5 million, which means that a risk of 1 in 100,000 should be prevented at a cost of $50.[9] This is the amount we would be willing to spend to save a statistical life immediately. The positivists argue that if a statistical life in two hundred years is worth the same amount, $5 million (in constant dollars), we should be willing to put aside only the present value of that amount to save that life. They are not discounting the lives—both are worth $5 million—but they discount the dollars, because a dollar put aside today grows with the discount rate.

That said, a positivist has nothing to contribute to the ethical debate about whether future lives and current lives should be given the same weight. Some positivists might think that people today actually do give less moral weight to people who live in the future than to people who live today; and that therefore a good positivist should recommend projects that reflect this belief. Other positivists might disagree. We will turn to this debate shortly.

Complications and Difficulties

This simple analysis runs into several complications.

1. *Private and Social Rates of Return.* A number of technical issues must be addressed in computing the opportunity cost. One

important issue turns on how to adjust for taxes and similar items, which cause a divergence between the rates of return investors see (the after-tax rate of return) and the rate of return that benefits society (the full, pre-tax rate of return because the investor gets the after-tax amount and the government gets the taxes). Another problem is that there are many market rates of interest. Treasuries pay a different rate from corporates; short-term bonds pay a different rate from long-term bonds; stocks and bonds have different returns. Also, discount rates can vary over time. While important, these issues are largely technical and need not detain us here.[10]

2. *Uncertainty.* A seemingly technical issue, very much worth our attention, involves the effect of uncertainty on discount rates. Suppose that we are considering a project that produces a return of $1 million in fifty years. Suppose too that the discount rate is uncertain and can take one of two values: 10 percent or 2 percent with equal probability. What is the expected discount rate we should use in evaluating this return?

It turns out not to be the simple average of 10 percent and 2 percent (i.e., 6 percent). Instead, the number is far lower—in this case, around 3.4 percent. The reason is that in order to determine the expected discount rate, we need to take the discounted value of the million dollars in each of the two circumstances and average these numbers. If the discount rate is 10 percent, the present value is about $8,500. If the discount rate is only 2 percent, the present value is $372,000. The average value is $190,000. This average, $190,000, is the number we should use for our estimate of the present value of the project. The implied discount rate (i.e., the discount rate that gives a present value of one million as $190,000) is 3.4 percent. As uncertainty increases and as the length of time increases, the effect is magnified.

This point has serious implications for the problem of climate change, which is the paradigmatic case of a long-term problem with uncertain effects.[11] Our conclusion means that the discount rate used by the positivists should be *near the very lowest expected rate of return over the long run.* Even if we fully expect growth rates over the next one hundred years to be 3 or 4 percent or more, there is a possibility that such rates will turn out to be very low. We should discount at the low rates because the bad states of the world—where growth is very low—will dominate the averaging process.

An important conclusion follows. We should not mistake the high observed rate of return on investments today to mean that positivists should recommend a high discount rate. On the contrary, the discount rate recommended by the positivists might be very low or even negative, especially because climate change might itself lower the rate of return on investments. This is true even if very bad climate consequences are unlikely because the averaging effect just illustrated means that bad outcomes, even if unlikely, dominate the analysis. We do not attempt to compute the resulting number here, but Stern uses a 1.4 percent discount rate (for entirely different reasons) and this may not be far off. Nordhaus's far higher number seems implausible given the mathematics of averaging over uncertainty.

3. *The Richer Future.* Positivists also have to be sure that they attach the correct values to items in the future.[12] If future people are richer than people alive today, they may value the environment more than people do today; it is a well-known fact that people value the environment more as they get richer. Moreover, if climate change damages the environment (as expected), the benefits it provides will be scarcer and its relative value increased. To say the least, estimating the correct values of the environment in the far distant future is not easy. But unless we are careful to take into account these sorts of considerations, we are at risk of using the wrong valuations and calculating the costs and benefits of climate change abatement incorrectly.

The Ethicists' Position

The ethicists argue that the only way to determine the correct discount rate is to go back to first principles of ethical reasoning.[13] Their central argument is that cost-benefit analysis with discounting can result in clearly unethical choices. Climate change provides the most vivid example. Climate change exposes the future to the risk of terrible harm. Because of the high discount rates required by the positivists' approach, however, we may be willing to spend only a very small amount today to prevent these serious harms in the future. If we respect a principle of intergenerational neutrality and believe that we have an obligation to take the interests of members of future generations seriously, this is unethical. Discounting cannot justify refusing to spend small amounts to prevent causing the risk of terrible harm to others.

To be more concrete, suppose that sea level rise will destroy Florida in two hundred years. Suppose also that these effects will be very difficult to counteract: reducing emissions may be extraordinarily expensive and mitigating the resulting harm not feasible. Under the positivists' approach, the rate of return on projects to reduce climate change is low, and cost-benefit analysis using the market return as a discount rate might recommend that very little be done. This conclusion, however, does not establish that we are behaving justly toward our descendants. We may, under a variety of ethical theories, owe them prevention of or compensation for this harm. Under a deontological approach, for example, the generation in which one finds oneself is irrelevant from the moral point of view; the current generation violates its moral obligations if it enriches itself while subjecting future generations to catastrophic harm. Under a welfarist approach, the question is whether the actions of the current generation increase or decrease overall welfare; it is easily imaginable that a failure to take certain actions to prevent climate change could cause a far greater welfare loss, to all generations taken as a whole, than would those actions themselves. The positivists' theory of discounting has nothing to say about the obligations of the current generations. In short, choosing projects solely through cost-benefit analysis with discounting can result in serious injustice and may violate our ethical obligations to the future.

By examining the details of discounting, the ethicists show why it produces what seem to be obviously unethical results. They offer three central reasons.

1. *Private versus Social Rates of Return.* The ethicists argue that the rate of return on an investment seen by individuals—the so-called private rate of return—is not the same as the benefit society gets from an investment—the social rate of return. There could be many reasons for this, but unless markets are perfectly competitive—a condition that is unlikely to hold—the two will not be equal. Observed interest rates reflect only private rates of return and, therefore, they

are not a good guide to whether the total benefits from an investment are worth the costs.

To illustrate using the discount rates discussed above, individuals may demand a 5.5 percent rate of return on investments, but they see only their private benefits. The social benefits of an investment may be much greater. It may, for example, be the case that if individuals get a 1.4 percent benefit that the additional benefits to society from an investment make it worthwhile. Looking only to the market rate, 5.5 percent, would mean that we reject projects that overall are worthwhile. Absent very stringent conditions, such as the economists' imaginary perfect market, the private and social benefits will not be equal, so we cannot look to the private returns available in the market as a guide for social policy.

2. *Individuals Neglecting Posterity.* Second (and closely related to the first argument), individuals determine today's interest rate by deciding how much to save for the future. In making this decision, individuals are considering their lifetime consumption—how much to consume today compared to their retirement—and possibly the consumption of their children or grandchildren. Climate change, however, is a problem that will last many hundreds of years, spanning multiple generations. Individuals, the ethicists claim, are simply not thinking about the distant future when making savings decisions. This means that observed interests rates are not a good guide for decisions over very long time periods. Individuals in setting the market rate of return are simply not considering the relevant question.

3. *Changing Rates of Return.* Finally, climate change abatement will involve large adjustments to the economy. If we make these adjustments, rates of return are likely to change. That is, the rate of return to investments is endogenous to the problem, not something that is external to the problem. It is a variable we choose rather than a variable we observe. To make this choice, we have to be explicit about the reasons. We have to go back to ethics.

• • •

To determine the appropriate discount rate, the ethicists imagine society conducting a thought experiment. Here the relevant claims become somewhat technical, and to understand them, we have to introduce a little math. Imagine that we are considering investing an additional dollar today that will produce returns in the future. As an ethical matter, how much *should* that return in the future be to compensate for the reduction of consumption today? The difference—how much the future should get if the present loses a dollar—is based on our ethical judgments about how much each generation deserves. The "social discount rate" reflects this ethical judgment. So if it is morally appropriate to give up 1 today, in order to obtain $1+\rho$ for the future, ρ, the Greek letter rho, is the social discount rate.

How should ρ be determined? There are three relevant variables. First, we might believe that the well-being of a future generation should receive more or less weight than the well-being of our generation, or the same weight. It is traditional to let the Greek letter delta, δ, equal the rate at which we discount the welfare of future generations, if at all, in the sense of discounting the moral importance of their well-being. This term is often called the "pure rate of time preference." Most ethicists believe that $\delta = 0$, that is, that we should give the same moral weight to our generation and future generations.

Formally, the optimal outcome is:

$$\text{Max } [W(C_0) + W(C_1)/(1 + \delta) + W(C_2)/(1 + \delta)^2 \dots]$$

where $W(C_i)$ is a measure of the welfare of a given generation i, C_i being the consumption of that generation. If $\delta = 0$, then one chooses the level of consumption and saving that maximizes the sum of the welfare of each generation.

Second, one needs to take into account the possibility that the marginal benefit from consumption is decreasing. Ethicists assume that the welfare of each generation is determined by a very particular form: as consumption goes up, the marginal benefit of additional consumption (in percentage terms) goes down at a constant rate, η, the Greek letter eta. In particular, the welfare of a given generation, W, is set equal to $C^{(1-\eta)}/(1-\eta)$. This is largely chosen for convenience rather than based on any empirical or ethical support—this functional form happens to be easy to work with mathematically.[14]

Suppose that growth rates are high, so that the future is much richer than today. Eta tells us that an increment to their consumption matters less because they are richer. The higher eta is, the more we demand the future get to give up a unit of consumption today. We can think of it as an inequality parameter.

Third, one must make an assumption about the growth rate of the economy. It is standard to use the symbol "c-dot," $\dot{c}$, to designate the rate of growth of the economy. (The mnemonic is that c is consumption and the dot indicates that the variable refers to the rate of change of consumption.) It tells us what the consumption in each period, the C's in the formula, will be.

With these assumptions, the ethicists derive the following as an expression for the social discount rate

$\rho = \eta\dot{c} + \delta$

As discussed, ρ is the social discount rate—that rate which we should use for evaluating projects such as climate change abatement. The term $\dot{c}$ is the growth rate. Eta reflects how much we care about inequality. The term δ is the pure rate of time preference.

Most of the discussion in the literature has been about δ. The ethicists almost uniformly take the position that it is unethical to allow δ to be positive because this would mean giving less weight to a future person simply because he lives in the future. If all people count equally, δ must be zero.

Eta is also important. It reflects views about inequality, both across generations and within any given generation. Higher etas are more egalitarian. In the climate change context, the more egalitarian we are, the higher the discount rate and the less we should be willing to invest in abatement. If we use the same values for redistribution with the current generation as across time, the more we want to redistribute now (i.e., we are more egalitarian), the higher the discount rate should be (we should want to engage in climate change abatement less because it redistributes toward the richer future). This seeming paradox, that strong egalitarians should care less about the future, arises because the future is expected (we hope) to be richer than today.

The terminology in the debate, at this point, becomes a little confusing. Setting δ equal to zero is often described as not discounting. In a sense this is correct: future generations are not discounted merely because they are in the future. The "pure rate of time preference" is zero. The social rate of discount, ρ, however, will be positive, but this, the ethicists say, is not because we are discounting. It is because we have distributive preferences: to the extent the future is richer, we should less want to increase their consumption. If ρ is positive, however, there will be a mathematical procedure used in evaluating climate change that is identical to discounting. The procedure, however, is about adjusting for distributive preferences, not time. It looks like discounting, say the ethicists, only by coincidence.

Finally, the social discount rate will reflect the fact that the economy is growing. The higher the growth rate, holding η constant, the higher that ρ will be. As we saw, this is what the positivists usually have in mind when they advocate "discounting." If the economy is growing, then putting money in a safe investment may help the future more than certain types of in-kind transfers like greenhouse gas abatement.

Once we have had our ethical debates about the parameters (and made good technical estimates of the growth rate, $\dot{c}$), we can determine the social discount rate. Stern in his initial report set $\delta = 0.1$, $\eta = 1$, and estimated the growth rate as 1.3 percent. Therefore, he used a social discount rate of 1.4 percent. Note that this is well below the rate of return available for investments in the economy. He has recently modified his views, adjusting η up to 2, so that the discount rate would now be 2.7 percent, closer to but still below the rate of return on other investments, although he has not recomputed the resulting policy recommendations.[15] Nordhaus would set $\delta = 1.5$, $\eta = 2$, and a growth rate of 2 percent, to derive a discount rate equal to 5.5 percent. The difference in conclusions from using these different discount rates is, as noted, dramatic, shifting the policy recommendation from among the most conservative to the most aggressive.

The Ethics of Discounting

Our Central Claim

In this section, we defend three conclusions. (1) The ethicists are correct to insist that choosing projects solely through cost-benefit analysis with discounting can result in serious injustice to the future and that we must respect the principle of intergenerational neutrality. (2) The positivists are correct that choosing any project that has a lower rate of return than the market rate of return throws away resources. (3) Notwithstanding the long debate between these two positions, they are not fundamentally inconsistent. The ethicists' insistence on intergenerational neutrality does not justify rejecting discounting at the market rate of return (properly taking uncertainty into account). As we shall soon see, choosing our overall legacy to the future—how much each generation gets—is to a large extent unrelated to the choice of particular projects.

In particular, the ethicists' concern over the effects of climate change suggests that we need to do more for the future, but it says nothing about what in particular that should be. The positivists' use of the market rate of discount to choose projects says nothing about whether we are fulfilling our obligations to the future.

To illustrate how the two issues can be separated, imagine that the current generation is leaving a particular legacy for the future, a legacy that for now we imagine to be the ethically justified amount. In monetary equivalents, call it $100. (Imagine putting as many zeroes at the end as needed.) Suppose also that a new project—say education or research into new technologies—is being considered that costs the current generation $10 and produces $20 for the future. If we engage in this project, we can reduce our legacy elsewhere, still leaving $100 for the future, maintaining distributional neutrality. The only question in this case is whether spending the $10 on this project produces a better return than spending the $10 elsewhere. The correct procedure for deciding whether to engage in this project is to measure the opportunity costs, which, as we showed above, is equivalent to discounting.

Now suppose we find out that our legacy to the future is inadequate because of newly discovered environmental harms from our actions. Suppose, for example, we discover that it is only $70 instead of $100. We must now reevaluate whether we are leaving enough. If we believe that, given this information, the correct amount to leave to the future is $95, we must increase our legacy.[16] We should do so in a way that costs us least, which means considering the opportunity costs of alternative projects. If we can find a project that costs us only $10 and leaves $25 for the future, we should not engage in projects that cost more.[17] The market rate of return measures the returns from currently available projects, so as an initial matter, the market rate is a measure of the opportunity costs of this choice. Once again, therefore, we should use discounting at the market rate to choose projects. Project choice and ethical obligations to the future are, to a large extent, separate.

Seen this way, the ethicists' criticism of the positivists' opportunity cost argument is simply irrelevant. It does not matter whether the current market rates of interest are ethically correct because they still represent the opportunity costs of investment. Recall the numbers used above: if we could invest $100 billion to produce $400 billion of benefits in one hundred years when that market rate is 5.5 percent, we could equivalently invest only $2 billion to produce those same benefits.

The ethicists argue that the 1.4 percent rate of return on the $100 billion (to produce $400 billion) is good enough once we consider the ethical arguments. That is, ethical considerations show that society should make investments not only with a 5.5 percent rate of return but also with a 1.4 percent rate of return. But given that there are hundreds or thousands of investment choices, if we are going to make investments at less than the 5.5 percent rate, we should start with those that have the highest return. If returns on education, public health, research into new technologies, or any number of other projects are higher than 1.4 percent, we should engage in these projects first. That is, at least initially, the opportunity cost is 5.5 percent. If we make enough investments to exhaust the opportunities at this rate of return, we can begin moving down the scale, but in no case should we jump to investments with very low rates of return as a first option.

Another way to describe the problem with the ethicists' approach is that if the correct social discount rate is 1.4 percent, we should be saving vastly more than we do today to leave the ethically correct legacy for the future. The exact implied savings rate is contested because the calculation involves estimates of technological change and other factors, but the ethicists seem to demand behavior of individuals that is far outside anything ever observed. As a first-order matter, therefore, the ethicists should not be arguing about discount rates for climate change. They should be arguing that overall savings and investment rates must dramatically increase. Given the dramatically increased savings and investment rates, we could decide whether investments in climate change abatement make sense.

This point is central: the ethicists' arguments are that we are leaving an insufficient amount for the future given current policies. On certain assumptions about the effects of current decisions on the future, these arguments may be correct. But regardless of whether they are, it says nothing about the particular choice of projects or policies. If we are going to increase the amount we leave for the future, it is incumbent on us not to do so in a way that wastes resources. Therefore, even if the ethicists' arguments are entirely correct, we still must carefully consider the opportunity costs of projects and pick those with the highest return.[18]

The positivists, however, also make a mistake. As we noted, using a market discount rate (properly adjusted for uncertainty) is not a reason for failing to discharge our obligations to the future. The underlying intuition behind the ethicists' argument is that current policies threaten to impoverish the future or to reduce greatly its welfare (as climate change threatens to do). If this is true, discounting is not a reason for allowing this to happen. It is simply a method of choosing projects that fulfill our obligation to prevent this from happening. A recommendation for modest climate change abatement, such as that made by Nordhaus, may also need to be accompanied by other projects that ensure the proper intergenerational distribution of welfare. That is, the ethicists may very well be correct that we need to adjust the amounts we are leaving for the future in light of our new understanding of the effects of climate change, while the positivists are correct that in doing so, we must be sure to pick those projects with the highest rates of return. Climate change abatement beyond the amount suggested by market-rate discounting would be justified if and only if it counts as such a project.

Objections

We consider here three objections to our claim that the positions of the ethicists and positivists address separate issues.

1. *The link between the market rate of return and overall savings rate.* The ethicists will object that our claimed separation between the market rate of return and our ethical obligations to the future is not correct. The reason is that if we were to save more, market rates would go down. When we finally are saving the right amount, the market rate will be the rate prescribed by their ethical arguments. Therefore, we might as well choose projects with that rate of return.[19] Similarly, as mentioned above, the ethicists might argue that the market rate of return is a choice variable, not an exogenous input, when we are dealing with large projects. The separation of the market rate of return and the choice of projects that we are suggesting does not make sense if the market rate of return is determined by the choice of projects.

While we agree that there is a likely connection between the overall savings rate or very large projects and the market rate of return, the ethicists' conclusion does not follow. If the market rate of return is, say, 5.5 percent and the ethicists argue that at the correct savings rate, the rate of return should be 1.4 percent, we should not immediately jump to projects with such low rates of return because eventually, if we increase savings enough, market rates of return might get this low. Large adjustments to our legacy to the future are difficult, and their success unclear. We should begin by choosing high-return projects, not low-return projects.

Moreover, it is probably wrong to suggest that if we sufficiently increase our savings, interest rates will be ethically appropriate—that is, will reflect the correct ethical weighting of future generations. Even the most sophisticated and modern models cannot compute equilibrium interest rates when there are large-scale changes to the economy, such as a vast increase in savings. It does not seem wise to make decisions by relying on the type of simplistic model used by the ethicists to discuss discount rates to argue for committing potentially trillions of dollars to a project on the theory that in the eventual equilibrium predicted by that model, the project choice will seem sensible.

A better decision procedure is to consider, more directly, the nature of our ethical obligations to the future. If the obligation is to leave more to the future than we previously thought (say because of the risks from climate change), we need to decide how to do that. The first choice of projects as we begin this adjustment should be those with the highest rate of return. This means discounting at the market rate. As market returns adjust (by going down if the model used by the ethicists is correct), the opportunity cost of new projects goes down.

2. *Feasibility.* While in theory the ethicists' and positivists' positions can be reconciled, the most difficult problems for both are potentially ones of feasibility. We start with the feasibility problems of the ethicists' position and then turn to the problems of the positivists' position.

The ethicists derive a discount rate to be used by a social planner independent of the rates demanded by individuals. That is, they start off with the basic premise that social rates of return and private rates of return are different and that the government should choose projects using the social rate of return. The problem with framing the question in this way is that individuals control most of the wealth in society. This means that whatever their preferences about savings, even if wrong, they can offset whatever the government does. Suppose, for example, that individuals, taken as a collective, want to leave $100 for the future and are doing so now. On the basis of the ethicists' recommendation, the government invests in a new project that leaves $40 for the future. Individuals seeing this project can simply reduce their legacy to $60 and keep the total at $100, frustrating the government's attempt to correct the market. If individuals can make these adjustments, the question the ethicists start with is simply the wrong question, because the government is not making the choice that is posed. Instead, the government is merely choosing which projects will be included in the total amount left for the future.

This type of behavior is known in the economics literature as Ricardian equivalence. The extent to which individuals behave this way is much debated and it is unlikely to be fully true.[20] Individuals may, under our numbers, reduce their legacy to $60 to completely offset the government project (the $60 individuals leave plus the $40 government project leaves a total of $100, exactly as before), or they may leave $70 or $80 or even $90, only partially offsetting the government project. Nevertheless, there is a basic futility problem with the ethicists' approach. The question they pose, by imagining that the government makes the basic choices about total investment, ignores the basic (and well-founded) constraints on the government.

The positivists also have feasibility problems. The problem is that it may not be possible to transfer resources across hundreds of years to compensate the future victims of climate change. We are, when it comes to climate change, like Robinson Crusoe—the choices are simply about distribution because there is no "bank"; there is no other way of shifting resources across these long periods of time.

This argument has been made most convincingly by Robert Lind.[21] He argues that we simply do not have direct methods of shifting resources across long periods of time, so the choice of projects such as climate change abatement have unavoidable distributional consequences. He imagines a proposal to transfer resources to the distant future through an investment with a 0 percent rate of return at a time when money or other projects earn a 10 percent return:

> The preferred decision may well be to make that investment and transfer the resources to the future generation even though it earns a zero rate of return. At this point an eager graduate student jumps up, sensing an economic slam dunk, and says "that was a really dumb decision. You could have invested that money at 10% and

made those people a lot better off." Wrong! We don't know how to set aside investment funds and to commit intervening generations to investing and reinvesting those funds for eventual delivery as consumer goods to the generation 200 years from now.[22]

The extent to which this is correct is an empirical and institutional question rather than a purely ethical question. We cannot rule out the possibility that projects with very low rates of return are the best way of shifting resources across time, but it seems unlikely. If, for example, the market rate of return were really 10 percent and the project at issue had a zero rate of return, it is hard to imagine that there were not other projects that, while perhaps not yielding the full 10 percent, had a higher rate of return than zero.

As in the quotation above, the claim that we cannot set aside funds for the future is often based on the problem of intervening generations. Suppose that a project, such as climate change abatement, will pay off in the long-distant future, say, one hundred to two hundred years. If we try to set aside funds for those same future individuals on the theory that the set aside funds will have a greater future value than funds invested in reducing emissions, they would have to pass through many generations before being received by their target. Any of those intervening generations could prevent the transfer to the future, making it impossible to guarantee that the funds will be used as intended.

Note, however, that the same problem arises with climate change abatement. Even if we spend vast resources reducing carbon emissions, future generations can always revert to burning fossil fuels. It is hard, without much more institutional detail, to understand why various projects would differentially have the problem of intervening generations, which is what is needed for the claim to have force. We cannot rule out this possibility, but it seems to us to be extremely unlikely.

Which way do the feasibility issues cut? We do not think that feasibility concerns seriously undermine the case for choosing the projects with the highest rates of return, which means using market discount rates. On the other hand, it seems clear that the very high savings rates suggested by the ethicists' approach are not realistic. At the same time, as we learn more about climate change, it is becoming apparent that our legacy to the future may be far less than we hoped. New understandings about climate change increase our need to save, whether through investments in abatement or otherwise. The ethicists' imperative is becoming more important. The best approach, however, is to help the public gain an understanding that our legacy is likely to be lower than we might have hoped and let the market aggregate preferences to the future with this new information incorporated.

3. *Incommensurable Goods: Lives versus Money.* A final argument that we cannot separate our ethical obligations and project choice is that climate change produces particular harms to the future that cannot be offset by other projects that help the future in other ways. This argument may take a variety of forms. One version is that the deaths caused by climate change cannot be offset by simply saving more. A prominent environmental academic, Dean Richard Revesz of NYU School of Law, makes such an argument.[23] He contends that the primary reasons for discounting monetary benefits do not apply to risks to life and health. Money is discounted because it can be invested. But human lives cannot be invested, and a life lost twenty or two hundred years in the future cannot be "recovered" by investing some sum in the present.

It is right to say that lives cannot be invested, but the problem with this argument is that what is being discounted is money, not lives. Under the standard analysis, any discount rate applies to willingness to pay to reduce statistical risks, which is a monetary measure.[24] The issues raised by the monetary valuation of lives is no different when used for standard cost-benefit analysis within a single time period and when used over differing time periods. Once lives (or, more properly, statistical risks of mortality) are converted to monetary equivalents, all of the arguments discussed above concerning discounting apply. If a life today and a life in two hundred years are both "worth" the same amount in terms of money, we need to discount the dollars allocated to the future life because money put aside for the future grows. For example, if a life today and a life in two hundred years are both worth $5 million, we should only allocate the present value of $5 million to the future life. Anything more than that would value the future life more than the present life. To be sure, translating lives into money is not easy and raises a host of thorny questions,[25] but it is not an issue for discounting in particular; objections to the methodologies for valuing lives are orthogonal to the discounting debate.[26]

An alternative version of the incommensurability argument, associated with the philospher Derek Parfit, is that even if discounting combined with changes in overall savings rates can get the allocation of resources correct across generations taken as a whole, harms to particular individuals cannot be offset by this procedure.[27] Parfit imagines activities today that increase the risk of genetic deformities in a small number of individuals in the future. Overall changes to the allocation of resources across generations do not compensate those individuals.

In a sense Parfit is right, but just as with the problem of valuing lives, this argument is not really about discounting. A project that takes place entirely within a single time period may still impose risks on individuals. Before the project is begun, all individuals may be subject to the same risk and taken as a whole the project may seem sensible. Ex post, however, particular individuals will suffer harm, and those who gain from the project may not be able to compensate those who lose. The arguments surrounding this problem have been debated vigorously.[28] It presents nothing new when the project occurs over more than one time period.

Climate Change and Respecting the Future

Our minimal goal here has been to illuminate the debate over discounting and intergenerational justice—to show exactly what is dividing the two sides. Because of the problem of uncertainty, we have argued for a low interest rate. But we have also contended that on the most fundamental question, the positivists are largely right: projects, including those involving climate change, should be evaluated by discounting the costs and benefits at the market rate of return, properly adjusted for uncertainty and for the value of the environment.

Any other approach risks choosing projects with low rates of return, leaving resources on the table.

Discounting, however, should be seen only as a method of choosing projects, not as a method of determining our ethical obligations to the future. We have endorsed a principle of intergenerational neutrality. Because of climate change, our legacy to our descendents appears to be far lower than we once thought. For this reason, we have a moral obligation to adjust. The proper response is to leave them more, not to choose projects by refusing to discount.

By now it should be clear that the discounting problem is just an intertemporal version of the distributive justice problem we discussed in chapter 4, and exactly the same mistakes are made in the two settings. Many advocates of a climate treaty that is generous (to poor countries, to the future) mix up ends and means. With respect to ends, it is ethically correct to give equal weight to a person who lives in our country, to a person who lives in a foreign country at the present time, and to a person who lives in our country or a foreign country in the future. As a practical matter, something less than equal weighting may be the best that we can achieve. Whatever the case, the key point is that these ends do not, by themselves, imply very much about the design of a climate treaty. As a first cut, the best way to help people in foreign countries is by distributing cash or in-kind foreign aid to them, today; and the best way to help the future is by saving and investing.

Now it may turn out that an appropriately designed climate treaty may include elements that redistribute more effectively to the poor today than foreign aid does; and, further, that it may include elements that redistribute more effectively to the future than other types of investment do (climate abatement schemes are just a type of investment). We are skeptical as to the first; we are agnostic as to the second. In any event, these conclusions require an empirical analysis that compares the rate of return of various alternative instruments.

Chapter 8

Global Welfare, Global Justice, and Climate Change

We have rejected the claim that climate change policies should be based on corrective justice or on an effort to redistribute from rich to poor. Moreover, the claim that emissions permits should be allocated on a per capita basis, while intuitively attractive, runs into many problems. Such an allocation is not easy to justify from the standpoint of any ethical theory, and efforts to insist on it may well derail a climate treaty, ensuring serious harms to poor people in poor nations. But what are the ethical obligations of wealthy nations?

We make four claims here. First, the moral worth of individuals transcends spatial and temporal boundaries. Wealthy people in rich nations have an obligation to help poor people, including poor people who live in developing countries. To the extent that climate change increases the differences between the rich and poor, this obligation increases. This point applies to temporal changes as well. If climate change would impoverish the future, the current generation has a strong obligation to take remedial action. If climate change would simply make some wealthy people in the future less wealthy than they would have been without it, but wealthier than most people today, then the obligation to take remedial action is weaker. And if climate change would produce future winners and losers, then the current generation's obligation to take remedial action today depends on the extent to which that action will in fact help the losers and not the winners, rather than vice versa.

Second, there is no requirement that this obligation should be met through a climate change treaty. Helping the poor requires complex judgments about what types of aid are best in a given set of circumstances. Those judgments depend on existing institutions and the behavior of other nations, including other powerful nations, other rich nations, and the governments of the poor nations themselves. It is not clear, for example, how best to discharge duties to the poor when they are ruled by corrupt governments or when powerful nations threaten or coerce them with bad intent. Indirect mechanisms, such as adjustments to the patent system or to trade or migration rules, may be more effective than direct mechanisms. A perfectly ethical nation, or even one with purely altruistic motives, would still have to make difficult choices about how best to behave in light of complexities of this sort. A climate change treaty is ill-suited for incorporating these sorts of considerations.

Third, the most important obligation with respect to climate change is to develop a broad, deep, and enforceable treaty that achieves appropriate climate goals. Nations must agree to join a climate treaty, which means that the treaty must be designed to make signatories better off. That is, a key for a climate treaty is what we have called International Paretianism—nations must believe that they are better off with a treaty than without. But the obligation to achieve a broad, deep, and enforceable treaty imposes a serious ethical duty on rich and poor nations alike—the obligation to cooperate. In our view, it is unethical for a nation to refuse to join a climate treaty in order to free-ride off of others. A nation might be better off with a climate treaty that it joins than it would be with no treaty, but it would be even better off if everyone else joined a climate treaty and it did not. Not joining a treaty in these circumstances, we will argue, is unethical.

Fourth, an effective climate treaty will likely generate a surplus, and the surplus must be distributed. We suggest that the surplus might best be used to reward states that have reduced emissions in the past. To the extent that a state took aggressive actions to reduce emissions, it should get a greater share of any surplus from a climate change treaty.

Each of these points would merit extended treatment. Here we merely sketch some of the most relevant considerations. We begin with a short discussion of the ethical backdrop before turning to each of the four claims.

Foundations: Welfarism and Deontology

We have referred throughout to two possible foundations for a climate change agreement. The first foundation is deontological.[1] Some people emphasize the moral irrelevance of national boundaries and urge that behind a "veil of ignorance," people would choose an agreement that is especially beneficial to those who are most disadvantaged. On a deontological approach of this kind, a climate change agreement might turn out to be justified if it improves the prospects of (say) the billion least-advantaged people on the planet even if it has significant adverse effects on the prospects of the billion who are less disadvantaged or most advantaged. Deontologists do not want to maximize welfare (a concept that they sometimes find mysterious). Instead, they seek to ensure that a climate change agreement is just, or fair, and many deontologists think that what is fair is what would be chosen by those who cannot know the nation in which they will find themselves. It seems intuitively plausible to say people would give special attention to raising the "floor," whether through cash transfers, emissions rights, or emissions reductions requirements that impose heavy burdens on wealthy nations and less heavy burdens on poor nations.

The second approach is welfarist.[2] Welfarists seek policies that maximize people's well-being, defined variously as their subjective sense of well-being, satisfaction of desires or preferences, or satisfaction of certain objective parameters.[3] They vary in how much they weight the welfare of the poor as compared to others, but almost all welfarists agree that giving a dollar to a poor person instead of a rich person improves welfare because the dollar provides more welfare for the poor person.

Welfarists sometimes borrow from the deontologists' standard repertoire and urge that behind the veil of ignorance, people would, in fact, choose welfarism. Some welfarists might maintain that if people did not know the nation in which they will find themselves, they would seek to improve their chances by maximizing the welfare of the average person on the planet. Alternatively, welfarists might abandon the idea of any veil of ignorance and simply urge that the best approach, and also the most just, is one that most increases the welfare of a relevant population. On a welfarist view, a climate change agreement that dramatically increases the welfare of all or most people in China and India (over a third of the world's total population) would be justified even if it slightly decreases the welfare of all or most people in the United States and Russia.

The debates between deontologists and welfarists raise some of the deepest questions in all of philosophy, and we do not attempt to resolve those debates here. There are obvious puzzles in deciding whether the best arguments for deontological approaches apply across national boundaries, and also about how, exactly, to specify the idea of welfare.[4] For our purposes, it is possible to bracket the largest debates, because the welfarist approach that we are inclined to support is broadly compatible with most deontological claims as well.

There is an orthogonal debate between cosmopolitans and their critics. Many deontologists are cosmopolitans, in the sense that they do not believe that national boundaries are relevant for questions of justice.[5] But others are not cosmopolitans: they believe that a national community is a prerequisite for justice, and that the relations between states are governed by more limited ethical constraints and are, in the main, pragmatic. Welfarists writing in the philosophical tradition tend to be cosmopolitans, but economists and others writing for public policy usually assume (without defense) that the relevant social welfare function should include only nationals. There may be practical reasons for making that assumption.

To the extent that ethical considerations require nations and individuals to act purely as cosmopolitans, as if there were no state system, climate policy would be much simpler.[6] Nations would be ethically obligated to enter into the optimal climate abatement system, such as an internationally harmonized carbon tax, and simultaneously adjust worldwide distribution (say through taxation and spending) to produce the optimal level of redistribution. Economists regularly study this scenario within the framework of welfare economics. They generally do not specify whether they are modeling a single nation or the entire world, but if we interpret their models to cover the entire world, governed by a single authority (or by nations acting solely in the interest of the entire world), the solutions they propose are consistent with a purely cosmopolitan approach. They generally prefer a carbon tax over other emissions control mechanisms and would couple such a tax with changes to other parts of the tax and spending system to offset any negative distributive consequences.

Many people, of course, do not believe that nations or people have significant obligations to other nations and to foreigners. They cannot help feeling more obligated to their own children than to children they have never met, and they extend this feeling to suggest that they have greater obligations to their fellow citizens than to foreigners. While we may discount such feelings as lacking moral standing, some people also believe that there are good reasons, founded in either principle or pragmatism, for having nations pursue the interests of their own citizens.

We do not mean to take a stand on the deepest questions here. However plausible cosmopolitan arguments might be in principle, they must come to terms with the fact that the world is divided into nations. Ethical arguments must ultimately balance global and local goals, feasibility within the state system, and maximization of global welfare (or the relevant deontological goals). The existence of nations has to be viewed as a basic constraint on ethical arguments.

Our goal is to identify ethical obligations that are broadly compatible with the core elements of both deontological and welfarist approaches, and that balance the practical realities and benefits of a state system with the cosmopolitan view that people around the globe matter, not just people in one's neighborhood or nation. This ecumenical approach means that we cannot make fine-grained distinctions based on the details of a particular theory. Instead, we must talk in generalities. Nevertheless, we think that our four basic claims should command general assent. Holders of differing ethical stances are likely to disagree about some of the details, such as the extent to which rich nations owe duties to the poor. Empirical matters, such how wealthy nations can best help the poor, are vigorously disputed, and certain judgments about the potentially catastrophic effect of climate change would affect some of our conclusions. We try to be clear on our assumptions and to steer a middle course through these disputes.

Duties of Wealthy Nations to the Poor

Our first claim is that wealthy nations have an ethical obligation to help the poor, including those living in other, poor nations. We made this claim in chapter 4, where we considered the obligations of rich nations to help prevent damage to a poor nation from a natural disaster such as an asteroid impact. Rich nations have an obligation in these circumstances to help poor nations, and this obligation goes beyond their natural charitable inclinations. We recognize that this claim is controversial, and we cannot defend it in detail here; but both of the leading approaches seem to converge on the basic conclusion.

On a welfarist view, nations have an obligation to take actions that improve overall well-being. Many people subscribe to some form of welfarism, including utilitarianism;[7] they believe that people are morally required to take steps that would increase overall welfare. Because a dollar would mean more in the hands of a poor individual than a rich person, helping the poor is likely to increase overall well-being. Nations also of course have a duty to help their own citizens: they must balance national goals with a duty to help the poor. Different people will balance these conflicting goals differently, but welfarists should generally support some sort of obligation to help the poor.

Deontologists might invoke Rawls's idea of a veil of ignorance: What principles of justice would reasonable people select if they were deprived of information about their own circumstances, including the nation in which they find themselves? Very plausibly, people would choose principles that would require rich countries to give a great deal to those in poor countries, especially if the latter are at serious risk. No less than race, sex, or disability, place of birth seems morally irrelevant; from the standpoint of justice, one does not have a right to a far better life because one was born in the United States rather than India. If this is so, wealthy people in wealthy nations have a moral obligation to provide assistance to poor people in poor ones, to reduce existing injustice.

On accounts of this kind, rich nations do not have large obligations to other slightly less rich nations. But wealthy nations have a significant obligation to those living in very poor nations; the extent of the duty to help the poor depends on the size of the differences in wealth or welfare. As we discussed in chapter 1, climate change threatens to make this difference greater because the worst impacts are likely to fall on poor nations. To this extent, the prospect of climate change increases the duties of the rich to the poor.

The Duty of Wealthy Nations to the Poor Need Not Be Discharged through a Climate Treaty

Our second claim is that there is no obligation to fulfill duties to the poor through a climate change treaty. Helping the poor is a difficult problem that depends on particulars about the context and local conditions. Nations need to consider these factors in determining how best to fulfill their obligations; they should not be bound to one particular policy, such as a redistributive climate change treaty. Moreover, a climate treaty is unlikely to be a good instrument for helping the poor. It is likely to be poorly targeted and expensive as compared to other mechanisms. We explored this issue in detail in chapter 4 and merely summarize the discussion here. There is both an ethical claim and an empirical component of this claim.

Ethical Claim

The ethical component is that duties to the poor should be considered in the context of the overall set of policies rather than for each policy. We should care about net transfers to the poor rather than whether particular individual policies, such as climate change policies, transfer resources to the poor. There is no obligation that each individual policy must be designed to achieve this goal.

In a sense, we can think of a climate policy designed to redistribute toward the poor as two separate policies: an optimal climate policy and a foreign aid program. The foreign aid program—the portion of the overall policy designed with distributive goals—has to be compared to other foreign aid programs. There are any number of other possible methods of making transfers. Wealthy nations choosing to transfer resources to the poor should do so in the most effective way. They should pick from all of the possibilities the set of methods that achieves their goal effectively. Distributive justice does not demand otherwise.

Empirical Claim

The empirical component of our claim is that reducing climate change (beyond the optimal amount) is likely to be an expensive way to help the poor. To be sure, reducing climate change is desirable in its own right and this is likely to help the poor as many of the potential victims are poor. But reducing climate change beyond the optimal amount or doing so in ways designed explicitly to help the poor, as was done in Kyoto, is unlikely to be the best method of helping the poor. As we have argued throughout, climate change is poorly targeted, whether one uses theories of distributive or corrective justice. Moreover, attempts such as Kyoto to adjust a climate treaty to take distributional considerations into account may significantly raise the cost and risk making the treaty ineffective.

There is an additional empirical constraint on using climate change to help the poor. Because national governments control the people who live in their territory, transfers of resources must almost always be to national governments rather than to the people themselves. Even if wealthy nations like the United States have a strong interest in advancing the well-being of poor people in poor states, and so would, in principle, be willing to agree to a climate treaty that had that effect, the viability of such a treaty would depend heavily on whether the governments of poor countries would actually use these transfers to help their poor. As we have emphasized, nations are not rich or poor; people are. All rich nations have poor people, and all poor nations have rich people. Experience with foreign aid teaches that simply transferring funds to a poor nation rarely helps the poor, in part because governments of many poor countries lack the institutional capacity or the will to help poor people.

It is both true and important that mitigating global climate change is an in-kind transfer that cannot be skimmed off by corrupt governments. But by the same token, that step is an extremely crude way to help the poor. Indeed, it does not help the poor today very much if at all; if anything, it will help the poor in the distant future. And it helps the rich as well as the poor; it is not well targeted. Finally, a realistic treaty, such as a cap-and-trade scheme, involves trusting these same corrupt governments to manage vast resources—billions of dollars of permits, for example—and to enforce rules against their wealthy cronies, such as the owners of greenhouse gas emitting factories. These problems may be surmountable, but they cannot be wished away, and if it is predictable that an ambitious climate scheme will simply enrich wealthy people and corrupt governments in poor states, and by the same token have little effect on the poor, then states should and will be reluctant to enter such a treaty.

There is a vast literature on development and the problems of helping the poor. It is fair to say that while there are many theories, right now we do not know how to best reduce poverty. Developed countries have spent trillions of dollars on the developing world with modest results. Many countries that have grown rapidly have received little aid from the developed world. Good local institutions and autonomy seem to be important, which significantly restricts what aid is able to do; it is not easy to impose markets or political institutions from the outside even if they are fundamental to growth. There is, moreover, no clear path to helping people who live in countries run by corrupt governments or that are ravaged by war.

Many methods of helping the poor might be subtle and indirect. Small changes can help build local institutions or solve important but local problems. Moreover, policy changes in other countries not directly related to poor countries can have important effects. For example, changes to intellectual property or trade or migration laws may be as important as aid. Climate change abatement may very well be part of this mix. The most we can say is that it is important for nations to continue to try to find ways to help the poor. We must not be constrained by considering separate policies with blinders, where each separate policy must independently meet distributive goals. Instead, getting good overall policies in place, including a good climate treaty, must be the goal.

A Global Climate Treaty

Our third claim concerns how to evaluate a climate treaty and the decisions of nations whether to join such a treaty. The argument here is brief but a bit dense; our basic emphasis is that nations have a moral obligation to cooperate rather than to free-ride.

The world needs to reduce emissions. While people disagree about the speed and extent of necessary reductions, it seems relatively clear that over the long run emissions from many sources, particularly from the energy sector, will have to decline dramatically, possibly to zero. Because doing so may significantly increase human welfare or because of deontological arguments for protecting the environment, the obligation to reduce emissions arguably rises to the level of an ethical obligation.

For reasons that we have sketched, a climate agreement must respect the principle of International Paretianism. This is not an ethical point, but it does lead to a conclusion that qualifies as such: If an agreement satisfies International Paretianism, it is unethical for a nation to free-ride on others by refusing to join such an agreement.

International Paretianism Redux

Effective climate policy needs global participation, at least in the sense that the significant emitters must be involved. The costs of the necessary emissions reductions go up dramatically when major emitting nations are excused from making reductions. Many of the lowest cost abatement opportunities are in developing countries. Emissions are growing at their fastest rate in developing countries. And a treaty that excludes developing countries is likely to be ineffective because of the opportunities for carbon leakage such a treaty would present. The central element in climate negotiations must be broad participation with all major emitting countries joining.

The only way to ensure broad participation is to design an agreement so that all nations are better off. A climate treaty must respect the principle of International Paretianism. This should be possible because the benefits from reducing emissions exceed the costs. Designing such a treaty will not be easy. It might require side payments or other transfers. Nations will bluff or hedge, claiming that their costs are high and benefits low. But if the benefits of emissions reductions truly exceed the costs, such a treaty should be possible.

When we use the term International Paretianism, we do not mean that states only pursue their narrow self-interest. Most people are altruistic to some extent, and it is in their interest to satisfy this altruism. For example, suppose that a climate treaty required the United States to lose, on net, say $20 billion, but that money would aid impoverished foreigners. To the extent that Americans are altruistic, the United States could consent to this treaty despite this apparent loss. We would say that such a treaty promotes America's self-interest.

Unfortunately, evidence of foreign aid flows suggests that most nations have only modest forms of altruism. Foreign aid is rarely more than 1 or 2 percent of GDP, and is given almost exclusively by the wealthiest nations. In addition, a substantial portion of foreign aid can be attributed to narrow interests rather than altruism. For example, if Japan or the United States gives foreign aid to a country in return for a favorable vote at a Security Council meeting, the aid cannot be said to reflect altruism; the sum of money is actually just a payment for a service, and is no more foreign aid than is payment for imported wheat. If we take into account such factors, the actual amount of foreign aid that reflects altruistic or moral considerations is extremely low. So by a roundabout route we are back to the position that International Paretianism, in practice rather than in theory, probably requires that all states that participate in a climate treaty are economically better off or at least not much worse off: that their abatement costs cannot be significantly more than the benefits in the form of avoided climate-related harms.

Other forces could cause a state's interest in a climate treaty, and hence its willingness to incur costs, to be greater than what would be predicted on the basis of a cost-benefit analysis that takes account only of economic interests. Suppose, for example, that a substantial number of people believe that they should make sacrifices to reduce climate change even when those sacrifices do not make them any wealthier. This belief might be based on a commitment to environmentalism or a concern about the well-being of people living in the distant future. They have a moral or ideological interest in limiting climate change, one that they would be willing to pay to see vindicated. If substantial numbers of people have such a belief, then a state's "interest" in reducing global greenhouse gas emissions would be greater, and thus it would be willing to incur treaty-related costs that would not otherwise be cost-justified. The extent to which people really feel this way, however, is unknown; perhaps international or intragenerational altruism will increase over time.

International Paretianism can be criticized from two perspectives. Idealists would argue that it wrongly rules out ethically desirable treaties. Realists would argue that it is too weak and optimistic, as it implies that states may consent to treaties that are not truly in their self-interest. Both these arguments deserve comment.

Idealists might argue that nations should not act in their self-interest—even in the broad sense that includes altruism—particularly when confronting a problem as difficult as climate change. And it is possible that ethical arguments will persuade people to set aside their self-interest and support a treaty that is morally proper. However, in the entire history of international relations, it is impossible to think of a treaty, based entirely or mainly on ethical ideals, that required sacrifices or transfers anywhere near the magnitude of those entailed by idealistic climate treaty proposals, such as the per capita system.[8] Could a climate treaty be the first such success? It could, but we believe that the risk of failure is too high to warrant experimenting with treaty proposals that would require the richest and most powerful nations to submit to unprecedented sacrifices.

Realists would argue that International Paretianism is too weak. If states act in their self-interest, then they will not only insist that the treaty improve their position relative to the status quo. They will also insist on the largest possible share of the surplus generated by the treaty. In effect, there is no surplus at all: the states with the greatest bargaining power—those states that gain the least from a treaty—will end up with the most limited obligations or with largest side payments. Their gains will not reflect ethical postulates but relative power. International Paretianism arbitrarily assumes that self-interested states take their self-interest only to a point, beyond which they submit to ethical demands. On this view, normative argument about a climate treaty is idle. A treaty will come into existence if and only if states benefit from a treaty, and the treaty will distribute benefits according to relative power.

The realist argument seems too strong, however. States might be swayed by ethical concerns; indeed, they might believe that acting consistently with ethical concerns will improve their reputation and in that way enhance their power. We cannot know for sure, but the possibility is realistic enough to justify ethical argument. The most important point made by the realists is that International Paretianism is arbitrary. If ethical concerns play some role, then states might agree to ethically proper treaties that do not serve their interests; if ethical concerns play no role or a very weak role, then even treaties that satisfy International Paretianism may not be possible.

International Paretianism is, at best, a rough attempt to solve the tension between realism and idealism. We see it as a pragmatic starting point for negotiations. Any treaty that deviates too far from International Paretianism has so little chance of success that it should not be the subject of negotiation. And while a treaty that complies with International Paretianism but has some ethical content may also be unattainable, any such treaty deserves serious consideration.

Free-Riding

International Paretianism, while necessary, is not sufficient to induce nations to sign a climate treaty. The reason is that even if a nation is better off with a treaty than without, it will be better off still if everyone else signs the treaty and it does not. This allows the nation to free-ride off of emissions reductions made elsewhere. Climate change is highly susceptible to free-riding because emissions reductions by one nation help everyone. Without a solution to the free-riding problem, a climate treaty cannot achieve broad participation even if it satisfies the principle of International Paretianism.[9]

Preventing free-riding will ultimately be a difficult problem of designing a treaty to encourage participation by reluctant nations. A climate treaty might have to be linked to other policies to effectively punish free-riders. To the extent that ethical concerns play a role in solving this problem, we believe that nations that refuse to sign a climate treaty so that they can free-ride off of emissions reductions by others are behaving unethically. Free-riding of this sort offends basic notions of fair play.

Deontologists have long supported such notions.[10] Rawls explores a case very much like that of climate change.[11] He considers a case where there is a mutually beneficial and just system of cooperation, which works only if everyone or nearly everyone cooperates, where cooperation requires sacrifice, and where noncooperating individuals cannot be excluded from the benefits. He argues that in this case, one may not accept the benefits without bearing the costs. The intuition is that you should not be able to exploit the cooperation and sacrifices of others for your own gain. The argument can be supported by simple intuitions about fairness or by arguments that consider the rules you would want to apply if you were randomly assigned to be either a cooperator or a potential free-rider.

Welfarists should support such an obligation on similar grounds. Without such an obligation, it is not clear whether or to what extent public goods, such as a climate change treaty, would be available. If there is an obligation to engage in actions that improve overall welfare, cooperating instead of free-riding, is obligatory.

In the climate context, the argument for a principle of fair play is even stronger than in the case Rawls considers. In the climate context, failure to cooperate imposes direct costs on other nations, either because there will be higher resulting concentrations of carbon and greater climate change or because the cooperating nations will have to spend additional resources to reduce their emissions. Opting out by a major emitting nation is very costly for other nations. So long as the climate treaty satisfies International Paretianism, free-riding in this case is unethical.

Dividing the Surplus

Suppose that the participation problem can be overcome and we are able to achieve a climate change agreement. If the benefits of reducing emissions are less than the costs, such an agreement will create some kind of surplus. How should that surplus be divided?

To understand what is at stake, imagine that the world contains two rather than 190+ countries. The states can jointly take more or less aggressive climate measures. More aggressive measures cost more; less aggressive measures cost less. More aggressive measures also create greater benefits in the form of avoided losses from climate change. The jointly optimal treaty will maximize the difference between the benefits and the costs.

If the states are rational, they will enter into a treaty that creates the optimal climate regime. Note that this is true even if the abatement costs for one state are greater than the climate benefits for that particular state: it will simply receive a cash transfer from the other state. But to keep things simple, let us assume that each state benefits in the absence of cash transfers. For example, Very Vulnerable State incurs a cost of 100 and obtains benefits of 900, while Pretty Vulnerable State incurs a cost of 400 and obtains benefits of 600. The costs are lost economic activity resulting from the higher cost of greenhouse gas intensive activities, while the benefits are avoiding disease, flooding, and so forth.

Very Vulnerable State and Pretty Vulnerable State face a classic bargaining problem. The treaty creates a surplus of 1,000 (800 + 200). Very Vulnerable State would be willing to enter a treaty that provides it with benefits of 100 or more, while Pretty Vulnerable State would be willing to enter a treaty that provides benefits of 400 or more. Under each of these treaties, neither state is made worse off because the benefits are greater than or equal to the costs. A number of alternative treaties would distribute the surplus more equally. A treaty can distribute the 1,000 surplus in many ways. For example, if the treaty creates a cap-and-trade system, it can allocate most of the permits to Very Vulnerable State, most to Pretty Vulnerable State, or any other share to each. Each will argue that it should have the larger share of permits. If the treaty establishes a uniform carbon tax, it can require side payments from one state to the other. Complicating matters, the distribution of the surplus can be hidden in other provisions, for example, those that require the sharing of technology.

But conceptually, the problem is straightforward. What is the best way to distribute the surplus between the countries?

The Baseline Case

One possible answer puts normative issues to the side and says simply that international politics will determine the division of the surplus. This is the realist view discussed above: feasibility doesn't just determine the baseline (that is, that states must be better off); it also determines the division of the surplus. In standard bargaining models, agents with greater bargaining power obtain the larger shares of assets that they are bargaining over. Bargaining power is usually understood to be some combination of sophistication and patience (low discounting).

In the bargaining literature, modelers sometimes assume that agents are more sophisticated if they are better educated, more intelligent, or more experienced. With a better understanding of the nature of the problem and the other side's weaknesses and strengths, the more sophisticated party ought to obtain a disproportionate share of the asset over which the parties bargain. More patient agents obtain a larger share of the surplus because (in standard bargaining games) failure to agree results in delay, and more patient parties are less injured by delay than less patient parties. Thus, to avoid costly delay, less patient parties yield a larger share of the surplus to more patient parties at the initial round of bargaining.

In the international setting, these observations suggest the unsurprising implication that larger, richer states will obtain the bigger shares of the surplus from a climate treaty. Large, wealthy states can afford to maintain large staffs of sophisticated diplomats, and they have more stable institutions, which cause agents to take greater account of the future. All of this suggests, not surprisingly, that industrialized nations and rising powers, including China, India, Brazil, and Russia, will largely determine how the surplus will be divided and will take a disproportionate share of it. On this view, there is no reason to believe that ethical norms will play a role in the division of the surplus.

However, our discussion above suggested that this conclusion is by no means obvious. It is certainly possible that ethical norms will play a role in bargaining. One reason is that public officials might be motivated by ethical norms even if those norms conflict with the demands of their populations. Agency costs give diplomats and leaders some latitude. We could easily imagine a situation in which the citizens of some nation do not much care about the welfare of people in other nations, but in which the leaders of that nation believe that it is important to take global welfare into account. Another and probably more important reason is that ethical norms may influence the attitudes of populations, who will demand that their leaders redress historical grievances, wealth imbalances, and so forth.

For these reasons, we will consider how ethical norms might play a role in the division of the surplus.

Normative Proposals

One possible approach would just reproduce the arguments that we criticized in earlier chapters. Even if a climate treaty must satisfy International Paretianism, this argument would hold, we can certainly divide the surplus in a manner that respects corrective justice or judgments about distributive justice. Corrective justice suggests that states that industrialized earlier (say, the United States) should receive a smaller portion of the surplus than states that industrialized later (say, South Korea). Distributive justice suggests that wealthier states should receive a smaller portion of the surplus than poorer states.

Although such a treaty would satisfy the feasibility constraint, some of our earlier objections would continue to hold. We criticized the corrective justice approach because of doubts about treating states as though they were individuals; about whether any state, or the individuals composing it, have been sufficiently culpable; about the difficult empirical questions, including the problem of toting up benefits as well as costs; and so on. These problems would need to be solved if corrective justice principles were used to divide the surplus.

We criticized the distributive justice approach because of concerns about whether rich nations should help future people rather than current poor people; about whether cash benefits might not be better than the in-kind benefit of reduced warming; about whether the governments of poor states would use cash transfers wisely; about whether it is good, if redistribution is our goal, to take steps that would penalize some poor people in rich countries and possibly benefit rich people in poor countries; and, most generally, about whether foreign aid is better accomplished in other ways. Not all the problems are the same. An optimal treaty would not redistribute excessively to wealthier people in future generations; all we are talking about here are cash (or emission permit) distributions that occur at the time that the treaty is ratified. We agree that, other things being equal, the surplus should be given to poor states rather than wealthy ones. But other things are not equal; incentive effects matter as well.

And we criticized reliance on the principle of equal division because this principle has limited normative appeal, is best justified as a tie-breaker when other principles run out, and, in the climate setting, has the highly counterintuitive implication that some poor and vulnerable states should bear a significant share of the cost.

There is a better approach. One could divide the surplus in a manner that rewards states that took aggressive measures to limit greenhouse gas emissions before the treaty was negotiated. The virtue of this approach is that it would help strengthen a precedent of international relations, to the effect that, when a global problem exists, states that act first to address it can expect to have their efforts rewarded, rather than ignored, when a treaty is ultimately negotiated. Such a precedent avoids a dangerously perverse effect: that states will delay responding to a looming crisis because early action would just weaken their bargaining position.

This approach bears a faint resemblance to corrective justice principles, but is not the same. Corrective justice focuses attention on states that have contributed the most to the current stock of greenhouse gas emissions. The alternative is to reward states that have recently undertaken efforts to invest in alternative energy sources, to cut back in other ways on greenhouse gas intensive activities, and to participate in good faith in climate-related treaty negotiations.

We suspect that all of these principles will play some role in the final negotiations. What is important to understand, however, is that they can at best play a modest role. At best, they can affect the distribution of the surplus; most likely, only a small portion of the surplus. It would be a mistake to use the negotiation of a greenhouse gas treaty as an opportunity to redress historical grievances and to correct global wealth inequalities. Such ambitious efforts will reduce the incentives of the most essential participants—wealthy, large states—to participate in negotiations. They might well distract from, and interfere with, the already formidable challenge of combating climate change.

There is another problem. The moral and practical factors that determine how the surplus will be divided are unlikely to coincide with any simple principle of justice or appealingly simple allocation scheme. Because nations will bring to the negotiating table different normative ideals, and because all nations, or at least all major nations, will need to agree on a climate pact, the distribution of the surplus will likely reflect different, possibly conflicting, ideals—not to mention different interests. This possibility spells trouble for simple allocation schemes—for example, the distribution of emissions permits on a per capita basis, which would require massive transfers from rich to poor countries, while having little relationship with any plausible moral ideal.

Pragmatism and Welfarism

However the surplus is divided, one should not lose sight of the fact that a climate treaty has a very strong ethical justification—it would promote the well-being of people around the world, and especially people in future generations. Even if the surplus merely reflects the bargaining power of states, the people in those states will be made better off, which is a good thing. Nearly all states, even very poor states, have bargaining power. If poor states are left out of a treaty, they can become havens for greenhouse gas emitting industry, and in the long run subvert much of the progress that would come from a treaty that included only powerful states as parties. For this reason, rich states will secure the consent of poor states to a treaty by offering them a share of the surplus rather than freezing them out. States driven entirely by their self-interest, including their interest in reducing global greenhouse gas emissions, will have strong pragmatic reasons for sharing the surplus with other states, even poor and vulnerable states, with the result that poor people will benefit. There is no reason to think that the resulting outcome will be ethically ideal, or even close, but there is strong reason to think the outcome will be better than the status quo.

Migration patterns and the vagaries of the future also play havoc with intuitions. If most of the gains from a climate treaty come in a hundred years or more, by then the descendants of many people in poor states today will live in rich states, either because their own state has finally developed or because they or their ancestors have migrated to a rich state. States tend to gain from each other's prosperity because wealthier people in richer states purchase goods from people in poor states and make investments in poor states. A climate treaty, by reducing the global cost of climate damage over the long term, will necessarily increase global wealth. Because the distant future configuration of the state system and the global economy is hard to predict, and indeed because the differential effects of climate change on different parts of the globe are still shrouded in mystery, negotiators today are effectively behind a veil of ignorance, and can do little to insure that the long-term benefits of the climate treaty will accrue to their own national group.

A Recapitulation

For many years, the principal disagreement over climate change has pitted the United States against Europe. To many people, especially in Europe, the United States has seemed to be the major obstacle to an international agreement. The dispute between the United States and Europe is not exactly over; European governments continue to seek more aggressive cuts than the United States does, and to be less insistent on various preconditions for an international treaty. But every year, that disagreement becomes decreasingly central, a kind of sideshow. The emerging division—and for the future, the most important one—is between the wealthy nations and the poor ones.

All around the world, wealthy nations are focusing on greenhouse gas emissions and expressing a willingness to reduce them. In sharp contrast, the developing countries have insisted that the principal duties lie with the developed world and that emissions reductions are a relatively low priority for them. As this book goes to press, newspapers report that officials in China, India, and other developing nations have made just this point. To a large extent, the difference between rich and poor nations reflects perceived self-interest—the constraints of International Paretianism. But much of the debate involves questions of justice. In this book, we have covered a great deal of material in a short space. It may be useful, by way of conclusion, to recapitulate the main lines of our argument.

Efforts to reduce greenhouse gas emissions will cost some nations far more than others, and benefit some nations far more than others. While projections remain highly uncertain, poor nations are much more vulnerable than wealthy nations, and India and Africa seem to be the most vulnerable of all. It is tempting to think that by virtue of their poverty, poor nations should not have to spend as much on emissions reductions as those that are rich. It might be added that to the extent that greenhouse gas emissions have been produced disproportionately by wealthy nations, the latter have a moral obligation to take the lead in solving the problem. And if and when an international agreement establishes emissions rights, it might seem clear that such rights should not be based on existing national emissions rates; per capita emissions rates seem to be a much more equitable foundation for the allocation.

We are in the unusual position of objecting to these claims while strongly supporting an international agreement to control greenhouse gas emissions and also favoring a great deal of redistribution from rich people in rich nations to poor people in poor nations—far more redistribution, in fact, than is being seriously proposed today. Our central argument has been that influential claims about climate change justice are both vulnerable in principle and dangerous in practice. As we have seen, it is not possible to defend an approach that requires wealthy nations to scale back their emissions while requiring poor nations to do little or nothing; such an approach would cost far too much and accomplish far too little. It would be somewhat better for the wealthy nations to agree to a treaty that requires more aggressive reductions than would promote their interests—if that treaty would deliver substantial benefits to poor nations, which are particularly vulnerable to the relevant risks. But such reductions are a crude method of helping poor people in poor nations; if that is really the goal, other steps such as foreign aid would probably be better.

It is tempting to think that rich countries should bear the principal economic burden of any climate change agreement, because they have been the major contributors in the past. But this claim encounters serious objections. As a matter of fact, developing nations will be close to the developed world, in cumulative emissions, in the relatively near future—perhaps by as early as 2030. As a matter of principle, the corrective justice model is a poor fit with the climate change problem. Many of the past contributors are dead. Many of them lacked the requisite state of mind. In addition, those who have inflicted risks of harm have also conferred significant benefits on other nations, and the benefits should be included in any accounting.

There is considerable intuitive appeal to the proposition that emissions rights should be allocated on a per capita basis. At first glance, it is not easy to defend the claim that the United States, with its 300 million people, should receive far greater emissions rights than China, with its 1.3 billion people. If redistribution is the goal, however, emissions permits should be allocated to poor countries, not on a per capita basis. We agree that little can be said, in principle, for allocating emissions rights in accordance with existing emissions rates. But no climate change agreement will be feasible if it amounts to a massive transfer of resources from wealthy nations to China and India. In addition, per capita allocations would create some unfortunate incentives.

Considerations of justice have an intertemporal dimension; they bear not only on the allocation of benefits and burdens across nations, but also on their allocation across time. We have supported a principle of intergenerational neutrality, which means that the citizens of later generations are entitled to the same weight as those of the current generation. But we have also argued that this principle does not justify a refusal to "discount" future costs and benefits. If climate change would impose significant harm in 2150, the current value of that harm is lower than it would be if it were anticipated to occur in 2100—and the current value of that harm is lower than it would be if it were anticipated to occur in 2050 or 2020. We have argued that future generations are generally helped, not hurt, by an approach that discounts future effects to present value. But we have agreed that under certain assumptions, cost-benefit analysis with discounting can lead to welfare losses and significant injustice, in violation of the principle of intergenerational neutrality. In such cases, the proper response is not to refuse to discount, but to take direct steps to ensure against welfare losses and injustice.

Our ultimate proposal has been for a forward-looking, welfarist approach to the problem of climate change that satisfies pragmatic constraints; one that takes the state system seriously by respecting International Paretianism and that refuses to entangle efforts to solve that problem with an assortment of other issues that admittedly deserve serious attention. Our own preference would be for a welfare-promoting agreement to deal with climate change, using science and economics to settle the optimal level of emissions reductions. To satisfy International Paretianism, the treaty may require side payments from states that have stronger interests in mitigating climate change to those that have weaker interests. States that have moved early to reduce their emissions should receive credit for doing so. On a separate track, states should continue to try to achieve redistributive or other justice-related goals, independent of climate-related issues. As we have emphasized, one of our central concerns is that an effort to achieve those goals in a climate agreement might prevent such an agreement from coming to fruition. It would be a cruel irony if the consequence of justice-related arguments were to doom the prospects for an international agreement—and thus to create exceedingly serious risks to human welfare, above all in poor nations.

Afterword: The Copenhagen Accord

From December 7 through 18, after this book was completed, delegates from 193 nations met in Copenhagen to discuss the next steps toward a climate treaty. Originally, it was thought that the Copenhagen conference would itself produce a climate treaty—a successor to the Kyoto Protocol—but long before the conference met nations made clear that they had not resolved enough of their differences for a treaty to be possible. Indeed, the Copenhagen conference turned out to be acrimonious and disorganized. Last-minute negotiations between the United States, China, India, South Africa, and Brazil produced a document known as the Copenhagen Accord, a nonbinding statement of principles. Other conference members did not formally sign the Accord, but instead "took note" of it in the final document emanating from the conference.

The Copenhagen Accord is a three-page document that states in general terms the principles that its signatories could agree to. The central principles are as follows:

- "Long-term" cooperative action must be taken in order to reduce emissions, with a goal of preventing global temperature from increasing by more than 2°C.
- Rich countries will provide financial support for adaptation efforts by poor countries. Developed countries will provide an amount "approaching" $30 billion for the three-year period, 2010–2012, and $100 billion per year by 2020.
- Non-Annex I parties to the Kyoto Protocol will "implement mitigation actions."
- Mitigation actions will be subject to "domestic measurement, reporting and verification" and communicated internationally.

- Nations will establish incentives for reducing deforestation and forest degradation.

It also appears that the further negotiations will be conducted mainly by the "major emitters"—about twenty countries, including the United States, China, India, the European countries, and Brazil. These countries collectively produce more than 90 percent of carbon emissions per year.

Measured against the goal of mitigating climate change, the Copenhagen conference was a fiasco. No treaty was agreed to. The principles are vague. The biggest emitters did not make any commitment to reduce their emissions by an amount sufficient to address climate change. China repulsed a proposal to provide for adequate monitoring of emissions so that commitments could be evaluated. The most ambitious countries—the European countries—were sidelined. Developing countries understandably insisted that they should receive more aid than developed countries were willing to provide, but, in the effort to get their way, they undermined negotiations and drove the major emitters to enter an agreement on their own.

The conference illustrates an important theme of this book. Nations have different preferences about the optimal extent of climate mitigation, and a successful climate treaty will need to ensure that those states with weak preferences will be made better off. Developing countries such as low-lying islands have the most to lose from climate change. At the same time, they have little bargaining power. Their demand that global temperatures be prevented from rising more than 1.5°C was brushed aside. European countries also have a great deal to lose from climate change, but, with their great wealth, they have the capacity to engage in significant cuts and to pay other countries to cut their emissions as well. The problem for the European countries is that they have little bargaining power. Their preference for reduction of emissions is well known, and they have already committed themselves through the Kyoto Accord. As a result, they do not have a credible threat to walk away from the negotiations if other countries fail to make concessions.

The United States and China remain the major players. The U.S. position—that, at least in the short term, it would reduce emissions to 17 percent below the 2005 level (far less ambitious than the Kyoto target for the United States)—was both weaker than necessary from a scientific perspective and, even then, not entirely credible to the rest of the world. President Obama could not commit the United States to reduce emissions because the power to do so is in the hands of the Senate, and public support in the United States for major emission reductions is weak.

China has offered to reduce its carbon intensity level 40–45 percent by 2020 from 2005 levels. (Carbon intensity refers to the amount of emissions produced per unit of GDP.) This offer, introduced with great fanfare, in fact would not commit China to take any additional mitigation measures. According to the Energy Information Administration's International Energy Outlook for 2009, China's carbon intensity will decline by this amount under business-as-usual projections (richer countries have more efficient energy sectors and so countries naturally emit less as they become richer). In 2006, China emitted 1,001 metric tons per million 2005 U.S. dollars of GDP; in 2020, China was projected to emit 558 metric tons. Moreover, a reduction in carbon intensity is not necessarily a reduction in emissions. Under the same projections, Chinese emissions will increase quite substantially (from 4.6 metric tons per person in 2006 to 6.6 metric tons per person by 2020). Further, China rejected the U.S. proposal to submit to international verification procedures, guaranteeing that any failure to reach its target would be invisible to the world, or at least contestable.

The lesson of Copenhagen is that a successful treaty will have to satisfy International Paretianism. International Paretianism requires nations to believe that they are better off with a treaty than without. For the governments of the rapidly industrializing states, economic growth takes precedence over climate change. China does not perceive it to be in its interests to agree to significant emissions cuts. Nor apparently do Russia, India, Brazil, and other major developing countries, which have so far refused to cut their emissions. We

suspect that a successful treaty will require transfers or other concessions to these nations, not to low-lying island states and poor countries in Africa. If the benefits of reducing emissions exceed the costs,
there should be a way to satisfy these constraints. But making transfers to countries that are increasingly perceived as geopolitical rivals
will face serious political and practical objections. And the target
constraint on global temperature increases will reflect a compromise
between countries that lose a great deal from climate change and
countries that lose less. Finding the solution to this problem should
be the central goal of climate policy.

International Paretianism also means that a climate treaty will not
include massive transfers to the poor to satisfy the demands of distributive or corrective justice. The conflict between poor and rich
threatens to overwhelm the essential goal of reducing emissions.
Poor countries initially demanded at least $100 to $200 billion per
year for adaptation costs. They deployed the corrective and redistributive justice arguments that we have discussed in this book. It
was the rejection of this demand that led to much of the chaos in the
Copenhagen conference. The poor countries greatly outnumber the
major emitters and have an equal vote under UN procedures. Thus,
they could easily use parliamentary maneuvers to block debate until
their demands were met.

In the end, the rich countries agreed to provide $10 billion per
year for the next three years, a figure significantly less generous than
what the poor countries demanded. Press reports surfaced that some
of this money would be raised by shuffling around foreign aid budgets. If true, poor countries will gain little or nothing from this commitment. The $100 billion per year figure starting in 2020 does not
reflect any clear commitments, as far as we have been able to ascer-
tain. The United States announced that it would contribute a share,
but did not specify the amount. The skepticism of poor countries is
easy to understand. Foreign aid is deeply unpopular in the United
States. The chief ambassador of the G77 group of developing nations
compared the Copenhagen Accord to "a suicide pact, an incinera-
tion pact in order to maintain the economic dominance of a few
countries. It's a solution based on values that funneled six million

people in Europe into furnaces." The words are an exaggeration but the sentiment is understandable. If climate change is to be addressed at a cost acceptable to people living in developed countries and the major developing countries, large-scale redistribution to the poor is not going to be a part of the climate treaty.

Copenhagen showed the futility of addressing poverty, past injustices, and climate change in a single negotiation. The problem of widespread poverty is urgent. So is the problem of reducing emissions. And many nations acted badly in the past. But no principle of justice requires that these problems be addressed simultaneously or multilaterally. Events in Copenhagen teach that trying to do so risks all three goals.

Notes

Introduction

1. There are few book-length treatments of the ethical issues connected to climate change. For recent and thoughtful treatments of the issue, see Steve Vanderheiden, *Atmospheric Justice* (2008); James Garvey, *The Ethics of Climate Change: Right and Wrong in a Warming World* (2008). A number of ethical studies are collected in *Fair Weather, Equity Concerns in Climate Change* (Ferenc L. Toth, ed., 1999). The reports by the International Panel on Climate Change contain discussions of ethical issues in a variety of places, most prominently in its Working Group III report, pages 142–47. Bert Metz, et al., *Climate Change 2007*, Mitigation of Climate Change, Working Group III Contribution to the Fourth Assessment Report of the International Panel on Climate Change. A number of analysts have considered how various ethical issues might be applied to a climate change agreement, analyzing, for example, how permits might be allocated across nations under various principles. Many of these studies are summarized in the IPCC report. The Stern Review considers ethical issues in its chapter 2 of Nicholas Stern, *The Economics of Climate Change* (2007). A great deal of work touches on ethical issues but is more oriented toward technical problems; see, for example, the essays collected in *Architectures for Agreement: Addressing Global Climate Change in the Post-Kyoto World* (Joseph E. Aldy and Robert N. Stavins, eds., 2007). Beyond these sources, there is a vast and unwieldy literature on the ethics of climate change—one that crosses numerous disciplinary boundaries. For a somewhat dated survey article of work by moral philosophers, see Stephen M. Gardiner, Ethics and Global Climate Change, 114 *Ethics* 555 (2004). Other earlier works include John Broome, *Counting the Cost of Global Warming* (1992); Donald Brown, *American Heat: Ethical Problems and the United States' Response to Global Warming* (2002); Michael Grubb, Seeking Fair Weather: Ethics and the International Debate on Climate Change, 71 *International Affairs* 463 (1995); Dale Jamieson, Global Responsibilities: Ethics, Public Health and Global Environmentall Change, 4 *Indiana Journal of Global Legal Studies* 99 (1998); Henry Shue, Subsistence Emissions and Luxury Emissions, 15 *Law and Policy* 39 (1993); Martino Traxler, Fair Chore

Division for Climate Change, 28 *Social Theory and Practice* 101 (2000); we cite some of this and other work in later chapters but we do not aim to be comprehensive.

2. Jonathan Wiener has made a similar point in a series of articles and books. See Jonathan Wiener and Richard Stewart, *Reconstructing Climate Policy: Beyond Kyoto* (2003); Jonathan Wiener, Global Environmental Regulation: Instrument Choice in Legal Context, 108 *Yale Law Journal* 677 (1999); and Jonathan Wiener, Climate Change Policy, and Policy Change in China, 55 *UCLA Law Review* 1805 (2008).

3. Sometimes ethical theories that focus on the effects of policies or actions on the welfare of individuals are called utilitarian rather than welfarist. Utilitarians think that we can measure the value of an act or policy based on how it changes the sum of the welfare of individuals, while welfarists more generally might think that other methods of aggregation are acceptable. The distinction will not play a role in our argument.

Chapter 1: Ethically Relevant Facts and Predictions

1. The three IPCC volumes from the Fourth Assessment Report, *Climate Change 2007*, alone contain several thousand pages summarizing the research in these areas. It is composed of Working Group I: The Physical Science Basis; Working Group II: Impacts, Adaptation and Vulnerability; and Working Group III: Mitigation of Climate Change, available at http://www.ipcc.ch.

2. IPCC Reports, *Climate Change 2007* (2007), available at www.ipcc.ch.

3. Eileen Claussen and Lisa McNeilly, Equity and Global Climate Change: The Complex Elements of Global Fairness, Pew Center on Global Climate Change (October 29, 1998); Marina Cazorla and Michael Toman, International Equity and Climate Change Policy: Resources for the Future, *Climate Issue Brief* 27 (December 2000); United Nations Environment Program, *Vital Climate Change Graphics* 14 (February 2005), available at grida.no/publications/bf/climate2; Ambuj D. Sagar, Wealth Responsibility, and Equity: Exploring an Allocation Framework for Global GHG Emissions, 45 *Climate Change* 511 (June 2000); Paul Baer, et al., Equity and Greenhouse Gas Responsibility, 289 *Science* 5488 (29 September 2000), available at http://www.climate-talks .net/2004-ENVRE130/PDF/20000929-Science-Equity.pdf.

4. World Resources Institute, Climate Analysis Indicator Tools (CAIT), available at http://cait.wri.org. Total GHG Emissions in 2000 with land use change (most recent land use data available).

5. Working Group I Report: The Physical Science Basis, IPCC Fourth Assessment Report 95, *Climate Change 2007* (2007); David Archer, *Global Warming: Understanding the Forecast, Part I: The Greenhouse Effect* (2007).

6. The strongest greenhouse gas is water vapor. Water vapor, however, cannot be changed directly by human activity for more than a very short period of time because the ability of the atmosphere to hold water vapor is determined by its temperature. If you boil a pot of water, vapor is emitted, but it will soon be rained out because, at a given temperature, the atmosphere holds only a fixed amount of water. Warmer temperatures mean more water in the atmosphere. The resulting increase in water vapor roughly doubles the warming from long-lived greenhouse gases.

7. IPCC WGI Report, supra n.5, at 100.

8. Except where specifically stated otherwise, all of the emissions data in this chapter are from the World Resources Institute, Climate Analysis Indicator Tool, available at www.cait.wri.org (accessed February 2009).

9. IPCC WGI Report, supra n.5, at 33.

10. IPCC WGI Report, supra n.5 at 283. The IPCC states that more than half of the CO_2 is removed from the atmosphere within a century, some fraction (20%) of emitted CO_2 remains in the atmosphere for many millennia.

11. It is possible that we might be able to increase greenhouse gas sinks.

12. Working Group III Report: Mitigation of Climate Change, IPCC Fourth Assessment Report 27, *Climate Change 2007* (2007), available at www .ipcc.ch/pdf/assessment-report/ar4/wg3/ar4-wg3-ts.pdf. (Lists 2004 global emissions at 49 gigatons, with significant uncertainty for emissions from methane, nitrous oxide, and from agriculture and forestry.)

13. IPCC WGI Report, supra n.5 at 37. An annual increase in the last decade of 1.9 ppm is reported.

14. James Hansen, et al., Global Temperature Change, 103 *Proceedings of the National Academy of Sciences* 14288 (September 26, 2006), available at www. pnas.org/content/103/39/14288.full.pdf. thtml.

15. Makiko Sato, GISS Surface Temperature Analysis, NASA Goddard Institute for Space Studies (2009), available at http://data.giss.nasa.gov/gistemp/graphs/.

16. IPCC WGI supra n.5, at 665.

17. David Archer, *The Long Thaw* 43 (2009).

18. Synthesis Report, IPCC Fourth Assessment Report 39, *Climate Change 2007* (2007), available at www.ipcc.ch/pdf/assessment-report/ar4/syr/ar4_ syr.pdf.

19. Reliability of Climate Change Models and Summary of Remaining Uncertainties, see IPCC WGI report, supra n.5, at 600, 666.

20. Nicholas Stern, *The Economics of Climate Change, The Stern Review*, (2006).

21. See Stern, supra n.20, at 98. The Stern Review itself only includes estimates for up to a 5-degree increase in global average surface temperatures. In

a more recent article, Stern extends the table to cover up to 7 degrees and we use these estimates.

22. See Martin L .Weitzman, On Modeling and Interpreting the Economics of Catastrophic Climate Change, 91 *Review of Economics and Statistics* 1 (2009).

23. Nebojsa Nakicenovic and Rob Swart, eds., IPCC Special Report on Emissions Scenarios (2000).

24. The IPCC is divided into three working groups: (i) science, (ii) impacts, and (iii) mitigation. When the IPCC issues a report, each of the working groups produces a volume covering their subject areas. The IPCC has now issued its fourth report. Working Group II of the Fourth Assessment Report is the volume in the fourth IPCC report that covers impacts.

25. There are a number of studies that attempt cross-model comparisons, including U.S. Climate Change Science Program, Synthesis and Assessment Product 2.1a; Scenarios of Greenhouse Gas Emissions and Atmospheric Concentrations (July 2007); and Energy Modeling Forum, The Cost of the Kyoto Protocol: A Multi-Model Evaluation, *The Energy Journal*, Special Issue 1 (1999).

26. Climate Change: Small Island Developing States, UNFCCC (2005); Working Group II: Impacts, Adaptation and Vulnerability, IPCC Fourth Assessment Report, *Climate Change 2007* (2007).

27. G. Yohe, E. Malone, A. Brenkert, M. Schlesinger, H. Meij, X. Xing, and D. Lee, *A Synthetic Assessment of the Global Distribution of Vulnerability to Climate Change from the IPCC Perspective that Reflects Exposure and Adaptive Capacity* (CIESIN [Center for International Earth Science Information Network], Columbia University 2006), available at http://ciesin.columbia.edu/data/climate/.

28. G. Yohe, et al., supra n. 27.

29. William D. Nordhaus and Joseph Boyer, *Warming the World* (2000).

30. Mark Lynas, *Six Degrees: Our Future on a Hotter Planet* (2008).

31. Ferenc Toth, State of the Art and Future Challenges for Integrated Environmental Assessment 4 *Integrated Assessment* 250 (2003).

32. For an interactive map showing the spread of malaria, see Impacts of Climate Change: Disease Spread, *Scientific American*; available at http://www.sciam.com/page.cfm?section=climate-map.

33. Richard S. J. Tol and Hadi Dowlatabadi, Vector-borne Diseases, Development and Climate Change, 2 *Integrated Assessment* 173 (2001).

34. Article 3, Section 1: Principles, UNFCCC, Full Text of the Convention, available at http://unfccc.int/essential_background/convention/background/items/1355.php.

35. Essential Background; Actors in the Negotiation Process, Article 3, Section 1; Article 4, Sections 1–3 and 9, UNFCCC, Full Text of the Convention,

available at http://unfccc.int/essential_background/feeling_the_heat/items/2915 .php.

36. See Stern, supra n.20, at 520. Stern provides more detailed calculations making this same point: broad participation is necessary to achieve reasonable climate goals.

37. See CAIT, supra n.4.

38. Hans-Werner Sinn, Public Policies against Global Warming: A Supply Side Approach, 15 *International Tax and Public Finance* 4 (2008).

39. Timothy Juliani, Senior Fellow at the Pew Center on Global Climate Change, Climate Change: Renewing U.S. Leadership in Challenging Times, Camden Conference (2009); Joyeeta Gupta and Michael Grubb, *Climate Change and European Leadership: A Sustainable Role for Europe?* (2000); President Barack Obama's speech to Governors Global Climate Summit promising global leadership on climate change (November 18, 2008).

40. See IPCC WGIII report, supra n.12, at chapter 6; Charles S. Pearson, *Economics and the Global Environment*, chapter 14 (2000); Peter Bohm, Cost-effectiveness and Facilitating Participation of Developing Countries in International Emissions Trading, Stockholm University (September 2001).

41. Zhong Xiang Zhang, Meeting the Kyoto Targets: The Importance of Developing Country Participation, 26 *Journal of Policy Modeling* 3 (2004).

42. A. Markandya and K. Halsnæs, *Developing Countries and Climate Change, The Economics of Climate Change* (A. Owen and N. Hanley, eds., 2004). The study looked at the results from sixteen different models, all addressing this question. When trading across regions is not permitted, the costs of meeting the Kyoto requirements range from around $200/ton for the United States to $400/ton for Japan, with the EU in the middle, at $305/ton. When trading is allowed within the developed countries, the average drops to $77/ton. If trading is allowed globally, the average drops to $36/ton. Brandt Stevens and Adam Rose (2002) have similar findings. Brandt Stevens and Adam Rose, A Dynamic Analysis of the Marketable Permits Approach to Global Warming Policy: A Comparison of Spatial and Temporal Flexibility, 44 *Journal of Environmental Economics and Management* 45 (2002).

43. See IPCC WGIII Report, supra n.12, at 33, 106.

44. See CAIT, supra n.4. They gather data from a variety of sources including the carbon inventories mandated by the UNFCCC, the International Energy Agency, and the World Bank. It is possible that the differences from the standard accounts that we find in the data are due to the CAIT data being somehow unusual, but this possibility seems unlikely.

45. World Bank Country Classification, available at http://web.worldbank .org/WBSITE/EXTERNAL/DATASTATISTICS/0,,contentMDK :20420458~menuPK:64133156~pagePK:64133150~piPK:64133175~the

SitePK:239419,00.html (accessed November 24, 2008). As of November 24, 2008, the World Bank classified all countries with per capita income of $11,456 or more as high income.

46. Russia is an Annex I country under the UN Framework Convention for Climate Change but is relatively poor, with per capita income of around $9,000 compared to the average for the EU 25 of more than $24,000 and a U.S. per capita income of more than $36,000. We do not include it as a wealthy country because its per capita income is below the World Bank standard for high income.

47. The data represent the additive sum of emissions over this time period, without adjusting for the possible decay of gases in the atmosphere. If data were adjusted for decay, earlier emissions would be weighted less. Although some of the details change when adjusting for decay, the basic conclusions do not.

48. The results once again do not change if we aggregate by wealth rather than by cumulative emissions. The top 40 wealthy nations have cumulatively emitted about 40 percent of the total.

49. Table 1.6 gives the simple sum of emissions from 1950 to 2000, without adjusting for the decay of carbon dioxide in the atmosphere. If we adjust the data for decay, the chart remains essentially unchanged except that the United States moves to eleventh, and Zambia takes its place in tenth place.

50. See IPCC WGIII Report, supra n.12. Figure 3a is an example of how aggregation can be used. That figure shows per capita emissions dominated by the United States and Canada, with Japan, Australia, and New Zealand plus various EU countries a distant second. Developed countries dominate the list of per capita emitters with the IPCC data. The WRI CAIT data come close to but do not exactly line up with the IPCC data when aggregated into similar regions. For example, the WRI data have the North American cumulative emissions up to 2000 at 19.23 percent, while the IPCC has the United States and Canada responsible for more than 25 percent in 2004.

Chapter 2: Policy Instruments

1. Lawrence H. Goulder and Ian H. Parry, Instrument Choice in Environmental Policy, 2 *Review of Environmental Economics and Policy* 152 (2008); Working Group III: Mitigation of Climate Change, IPCC Fourth Assessment Report, chapter 13, *Climate Change 2007* (2007); *Handbook of Environmental Economics*, chapter 10 (2003); Nicholas Stern, *The Economics of Climate Change: The Stern Review*, Parts 4, 5 (2007).

2. Jonathan Wiener, Global Environmental Regulation: Instrument Choice in Legal Context, 108 *Yale Law Journal* 677 (1999); Jonathan Wiener and Richard B. Stewart, *Reconstructing Climate Policy: Beyond Kyoto* (2003). Wiener and Stewart argue that permits allow us to pay off China but taxes do not.

3. There is a related argument that this same flexibility makes it easier to achieve an agreement with cap-and-trade regimes because it makes it easier to buy off recalcitrant nations. We do not address this argument here.

4. Charles Kolstad and Michael Toman, The Economics of Climate Policy, 3 *Handbook of Environmental Economics* 1561 (Karl Goran-Maler, ed., 2005); Garrett Hardin, The Tragedy of Commons, 13 *Science* 3859 (December 1968); J. B. Ruhn, et al., The Tragedy of Ecosystem Services, 58 *Bioscience* 969 (2008).

5. The exact cost savings are difficult to compute because they depend on a counterfactual assumption about what the costs would have been under an alternative regulatory regime. See Curtis Carlson, Dallas Burtraw, Maureen Cropper, and Karen Palmer, Sulfur Dioxide Control by Electric Utilities: What Are the Gains from Trade? 108 *Journal of Political Economy* 1292 (2000) for a discussion.

6. Gilbert Metcalf, A Proposal for a U.S. Carbon Tax Swap: An Equitable Tax Reform to Address Global Climate Change, Discussion Paper 2007-12, Hamilton Project, Brookings Institution (November 2007). A very rough estimate for the United States is that a $15/ton carbon tax would raise about $100 billion per year. To get a sense of this number, our total federal tax revenues were about $2.4 trillion in 2006, so the carbon tax would be about 4 percent of total federal taxes. Corporate tax receipts were about $350 billion in 2006, so a carbon tax would be more than one-quarter of the corporate tax base. Carbon tax revenues would easily exceed revenues from the estate and gift tax.

7. See Gilbert Metcalf, Federal Tax Policy Toward Energy, 21 *Tax Policy and the Economy* 145 (2007).

8. For a discussion, see Michael Wara, Measuring the Clean Development Mechanism's Performance and Potential, 55 *UCLA Law Review* 1759 (2008); UNFCCC, Clean Development Mechanism In Brief (2008); Nancy Kete, Designing the Clean Development Mechanism to Meet the Needs of a Broad Range of Interests, World Resources Institute (2000).

9. Nicholas Stern, The Economics of Climate Change, 98 *American Economic Review Papers and Proceedings* 536 (2008). Many others make similar claims. See, e.g., Ross Garnaut, *The Garnaut Climate Change Review* 197 (2008).

10. FY 2008 International Affairs Congressional Budget Justification, U.S. 2008 Budget, available at http://www.state.gov/f/releases/iab/c21508.htm.

11. It is possible that entities within countries would be allocated the permits, but this seems unlikely given that the countries themselves would have to agree to the treaty.

12. The identical analysis applies with respect to the issue of auctioning compared to giving away permits. Permit systems have no greater flexibility in this regard as taxes can be designed to mimic any combination of auctions and free allocation of permits by exempting a given amount of emissions from taxation.

13. A carbon offset is a voluntary payment made to reduce emissions so as to offset emissions from normal activities. For example, if an individual is going to fly to a destination, he can purchase an offset that promises to reduce emissions elsewhere in an amount equal to the emissions from flying.

14. William Chameides and Michel Oppenheimer, Carbon Trading over Taxes, 315 *Science* 1670 (March 23, 2007).

15. Nick Johnstone, Efficient and Effective Use of Tradeable Permits in Combination with other Policy Instruments, in Greenhouse Gas Emissions Trading and Project-based Mechanisms, Proceedings of the OECD Global Forum on Sustainable Development: Emissions Trading GATEP Country Forum (2003).

16. Cedric Philibert, Price Caps and Price Floors in Climate Policy: A Quantitative Assessment, IEA Information Paper (OECD/IEA, December 2008).

17. Timothy Lenton, et al., Tipping Elements in the Earth's Climate System, 105 *Proceedings of the National Academy of Sciences* 1786 (February 12, 2008).

18. William Pizer, Climate Change Catastrophes, Resources for the Future, Discussion Paper 03-31 (May 2003).

19. See Geoffrey Heal and Bengt Kristom, National Income and the Environment, in 3 *Handbook of Environmental Economics*, chapter 22 (Karl-Goran Maler and Jeffrey Vincent, eds., 2005).

20. Similar arguments apply to carbon offsets. See http://www.cheatneutral .com/.

21. See Elizabeth Anderson, *Value in Ethics and Economics* (1995).

Chapter 3: Symbols, Not Substance

1. Our account of the Kyoto Protocol and its reception follows Cass R. Sunstein, Of Montreal and Kyoto: A Tale of Two Protocols, 31 *Harvard Environmental Law Review* 1 (2007).

2. Miranda A. Schreurs and Yves Tiberghien, Multi-Level Reinforcement: Explaining European Union Leadership in Climate Change Mitigation, 7 *Global Environmental Policy* 19 (2007).

3. See William D. Nordhaus and Joseph Boyer, *Warming the World: Economic Models of Global Warming*, chapter 8 (2000).

4. Senate Resolution 98, available at http://thomas.loc.gov/cgi-bin/ query/D?c105:3:./temp/~c105Fhqpms.

5. We follow here the analysis of Sunstein, supra n. 1, although we update the figures.

6. Table 2.1 of Annual European Community Greenhouse Gas Inventory 1990–2006 and Inventory Report 2008, EEA, available at http://www.eea .europa.eu/publications/technical_report_2008_6.

7. Kathryn Harrison and Lisa McIntosh Sundstrom, The Comparative Politics of Climate Change, 7 *Global Environmental Policy* 4, 13 (2007).

8. See UNFCCC, National greenhouse gas inventory data for the period 1990–2005, at 9 (October 24, 2007).

9. The European Union Emissions Trading Scheme, Insights and Opportunities, pp. 2, 7, Pew Center on Global Climate Change, available at http://www.pewclimate.org/docUploads/EU ETS%20White%20Paper.pdf.

10. Ibid.

11. Questions and Answers on Emissions Trading and National Allocation Plans (2005), available at http://europa.eu/rapid/pressReleasesAction.do?reference=MEMO/05/84&format=HTML&aged=1&language=EN&gui Language=en.

12. A. Danny Ellerman and Barbara K. Buchner, The European Union Emissions Trading Scheme: Origins, Allocation, and Early Results, 1 *Review of Environmental Economics and Policy* 66, 72 n. 9 (2007).

13. Emissions Trading: Strong Compliance in 2006, Emissions Decoupled from Economic Growth (2007), available at http://europa.eu/rapid/pressReleases Action.do?reference=IP/07/776&format=HTML&aged=0&language=EN& guiLanguage=en.

14. Emissions Trading: 2007 Verified Emissions from EU ETS Businesses (2008), available at http://europa.eu/rapid/pressReleasesAction.do?reference= IP/08/787&format=HTML&aged=0&language=EN&guiLanguage=en.

15. The European Carbon Market in Action: Lessons from the First Trading Period 12 (2008), available at http://www.caissedesdepots.fr/IMG/pdf_08-03-25_interim_report_en.pdf.

16. Ibid. at 17.

17. Ibid. at 15.

18. Ibid.

19. Proposal for a Directive of the European Parliament and of the Council amending Directive 2003/87/EC so as to improve and extend the greenhouse gas emission allowance trading system of the Community 7 (2008), available at http://ec.europa.eu/environment/climat/emission/pdf/com_2008_16_en.pdf.

20. Ibid. at 4.

21. Ibid. at 21.

22. Ibid. at 8.

23. For a critical discussion of this and related lawsuits, see Eric A. Posner, Climate Change and International Human Rights Litigation: A Critical Appraisal, 155 *University of Pennsylvania Law Review* 1925 (2007); for a more optimistic assessment, see Jonathan Zasloff, The Judicial Carbon Tax: Reconstructing Public Nuisance and Climate Change, 55 *UCLA Law Review* 1827 (2008).

24. An important case decided just as this book went to press held that the political question doctrine does not apply to climate suits. See *Connecticut v. American Electric Power Company, Inc.*, 2009 WL 2996729 (2nd Cir., 2009).

25. Congressional Budget Office, *The Estimated Costs to Households From the Cap-and-Trade Provisions of H.R. 2454*, Congressional Budget Office (June 19, 2009), http://energycommerce.house.gov/Press_111/20090620/cbowaxman markey.pdf.

26. U.S. House of Representatives, Committee on Energy and Commerce, *Committee Releases Updated Summary of American Clean Energy and Security Act* (June 2, 2009), http://energycommercehouse.gov/index.php?option=com_content&view=article&id=1635:committee-releases-updated-summary-of-american-clean-energy-and-security-act&catid=122:media-advisories&Itemid=55.

Chapter 4: Climate Change and Distributive Justice

1. See the various proposals by Brazil and other developing countries, collected in Daniel Bodansky, International Climate Efforts Beyond 2012: A Survey of Approaches, *Pew Center on Global Climate Change* (2000).

2. For a critical discussion of such claims, see Christopher Stone, Common But Differentiated Responsibilities in International Law, 98 *American Journal of International Law* 276 (2004).

3. Not all of the Annex I countries are wealthy, although most countries subject to significant restrictions under the Kyoto Protocol are.

4. Peter Singer, *One World: The Ethics of Globalization* 40 (2002); see also Steve Vanderheiden, *Atmospheric Justice: A Political Theory of Climate Change* 78 (2008).

5. Our argument is similar to the standard arguments about in-kind redistribution within domestic policy. For example, should we give the poor food stamps and housing vouchers, or should we have a more progressive tax system? While current policy takes a wide variety of approaches, the general answer, in our view, is that cash is to be preferred except in specific circumstances. In-kind benefits, for example, might allow sorting of those truly in need from pretenders, or they may be justified on paternalistic grounds. Essentially the same arguments apply in the international climate change context except that redistribution in the international context is, if anything, more complex and difficult than in the purely domestic context.

6. Martha Nussbaum, *Frontiers of Justice* (2005); Vanderheiden, supra n.4.

7. Nussbaum, supra n.6. Some people appear to believe that poor nations have an entitlement to help from wealthy nations. But even if this is so, assistance in the case we are describing is less valuable than direct financial aid—a point that we shall be emphasizing.

8. See, e.g., Joseph E. Stiglitz, *Economics of the Public Sector* 55, 92 (1986). Economists have long criticized such in-kind programs as paternalistic, and as less likely to be in the interest of beneficiaries than cash transfers. And although

a case can be made for paternalism when governments attempt to aid citizens who suffer from self-control problems, or are poor and uneducated, this case is far weaker, and provokes politically explosive memories of rationalizations of imperialism, in the context of government-to-government foreign aid.

9. Thomas Schelling, What Makes Greenhouse Sense, *Foreign Affairs* (May/June 2002).

10. We are putting to one side the possibility that technological change will make it easier to divert the asteroid in the future. By hypothesis, specialists believe that cost-benefit analysis justifies immediate action. But it is possible that because of technological advances, future generations will be able to eliminate the threat more cheaply than present generations can.

11. It might be the case that rich countries have greater abatement opportunities because they have access to better technology or have a greater institutional capacity to monitor emissions or some similar factor. These factors arise because rich countries are rich, but we mean here having rich countries go beyond the otherwise efficient abatement because of distributive concerns.

12. The Clean Development Mechanism is an important attempt to extend reductions to developing countries and reduce the problems with Kyoto that we discuss below. As we discussed in chapter 2, however, because the CDM is structured as a subsidy, it suffers from serious implementation problems. Essentially the same criticisms we make of Kyoto apply to the CDM: an inferior climate policy was chosen for distributional reasons.

13. See Schelling, supra n.9.

14. This point is a basic application of the idea that efficient policies along with redistribution are superior to inefficient policies that produce the same distributive outcome. See Anthony Atkinson and Joseph Stiglitz, The Design of the Tax Structure—Direct versus Indirect Taxation, 6 *Journal of Public Economics* 55 (1976). Others have argued that this point does not carry through to the climate setting. See Graciela Chichilnisky and Geoffrey Heal, Who Should Abate Carbon Emissions?: An International Viewpoint, 44 *Economic Letters* 443 (1994).

15. William Easterly, *The White Man's Burden: Why the West's Efforts to Aid the Rest Have Done So Much Ill and So Little Good* (2006).

16. Martin Weitzman, On Modeling and Interpreting the Economics of Catastrophic Climate Change, 91 *Review of Economics and Statistics* 1 (2009).

17. See Easterly, supra n.15.

18. Ibid.

19. William Nordhaus, *A Question of Balance* 159 (2008).

Chapter 5: Punishing the Wrongdoers

1. See, e.g., Jiahua Pan, *Common but Differentiated Commitments: A Practical Approach to Engaging Large Developing Emitters Under L20* (2004), available

at http://www.l20.org/publications/6_5c_climate_pan1.pdf; Peter Singer, *One World: The Ethics of Globalization* (2002); Steve Vanderheiden, *Atmospheric Justice: A Political Theory of Climate Change* (2008).

2. We do not address whether there are *legal* challenges, specifically tort challenges, to greenhouse gas emissions. There is an extensive literature on this topic. See, e.g., Eduardo M. Penalver, Acts of God or Toxic Torts? Applying Tort Principles to the Problem of Climate Change, 38 *Natural Resources Journal* 563 (1998); David A. Grossman, Warming Up to a Not-So-Radical Idea: Tort-Based Climate Change Litigation, 28 *Columbia Journal of Environmental Law* 1 (2003); David Hunter and James Salzman, Negligence in the Air: The Duty of Care in Climate Change Litigation, 155 *University of Pennsylvania Law Review* 1741 (2007). For a discussion of the possibility of tort claims brought under the Alien Tort Statute, see Eric A. Posner, Climate Change and International Human Rights Litigation: A Critical Appraisal, 155 *University of Pennsylvania Law Review* 1925 (2007). However, the tort claim and the moral claim are overlapping.

3. Calculated from 1950, excluding land use change and excluding greenhouse gases other than carbon dioxide.

4. See, e.g., Daryl J. Levinson, Collective Sanctions, 56 *Stanford Law Review* 345 (2003); Thomas Miceli and Kathleen Segerson, Punishing the Innocent along with the Guilty: The Economics of Individual versus Group Punishment, 36 *Journal of Legal Studies* 81 (2007).

5. Louis Kaplow and Steven Shavell, *Fairness Versus Welfare* 12 (2005). For this reason, corrective justice claims will not be appealing to welfarists, who tend to think that corrective justice is relevant, if at all, because it serves as a proxy for what welfarism requires. We tend to think that welfarists are generally correct here but bracket that point and the associated complexities for purposes of discussion.

6. See, e.g., Daniel A. Farber, Basic Compensation for Victims of Climate Change, 155 *University of Pennsylvania Law Review* 1605, 1641 (2007).

7. U.S. Census Bureau, 2008 Populatiton Estimates, T6-2008, available at factfinder.census.gov.

8. See, e.g., H. D. Lewis, Collective Responsibility, in *Collective Responsibility: Five Decades of Debate in Theoretical and Applied Ethics* 17 (Larry May, and Stacey Hoffman, eds., 1991).

9. Stephen Kershnar, The Inheritance-Based Claim to Reparations, 8 *Legal Theory* 243, 266 (2002) (describing and criticizing these arguments). These arguments are often analogized to unjust enrichment arguments. See Eric A. Posner and Adrian Vermeule, Reparations for Slavery and Other Historical Injustices, 103 *Columbia Law Review* 689, 698 (2003).

10. Posner and Vermeule, supra n.9, at 699.

11. In recent years, some philosophers have challenged traditional criticisms of collective responsibility, but these philosophers tend to ground collective

responsibility in individual failures to act when action was possible and likely to be effective, and when the person in question knew or should have known that she could have prevented the harm. See, e.g., Larry May, *Sharing Responsibility* 1 (1992); cf. Brent Fisse and John Braithwaite, *Corporations, Crime and Accountability* 50 (1993); Christopher Kutz, *Complicity: Ethics and Law for a Collective Age* 166 (2000); David Copp, Responsibility for Collective Inaction, *Journal of Social Philosophy* 71 (Fall 1991). These arguments do not carry over to the greenhouse gas case.

12. We might also think that Americans of, say, the last decade or two can be held responsible for their greenhouse gas emissions; because most of them are alive today, they might be considered obliged to provide a remedy.

13. Michael Saks and Peter Blanck, Justice Improved: The Unrecognized Benefits of Aggregation and Sampling in the Trial of Mass Torts, 44 *Stanford Law Review* 815 (1992).

14. For more on the causation problem, see generally Posner, supra n.2.

15. Stephen R. Perry, Loss, Agency, and Responsibility for Outcomes: Three Conceptions of Corrective Justice, in *Tort Theory* 24 (Ken Cooper-Stephenson and Elaine Gibson, eds., 1993).

16. Matthew D. Adler, Corrective Justice and Liability for Global Warming, 155 *University of Pennsylvania Law Review* 1859 (2007).

17. See Singer, supra n.1, at 43–9.

18. One commentator suggests 1990 as a date for when emitting activities could have become negligent. See Pan, supra n.1, at 3–7. We put to one side the following questions: What if a consensus did not exist, but many experts believed that climate change was likely and that if it occurred, the damages would be massive? How should negligence be analyzed if there were (say) a 30 percent chance of enormous harm? In principle, the benefit-cost test might find negligence in such circumstances (if the discounted harm exceeded the cost of precautions).

19. See William Nordhaus, *A Question of Balance* (2008).

20. Nordhaus, supra n.19, at 30; *EPA, Emission Facts: Average Carbon Dioxide Emissions Resulting from Gasoline and Diesel Fuel*, available at http://epa.gov/oms/climate/420f05001.htm. The figures in the text are very rough and are used for illustration only: what we say would be true even if the numbers are higher or lower, as long as they are not zero.

21. Matthew Adler makes this point in criticizing Farber's corrective justice argument; see Farber, supra n.6; see Adler, supra n.16. However, we disagree with Adler's argument that corrective justice can justify government-to-government claims; see ibid. at 1866, for reasons given below.

22. A vigorous argument in favor of such engagement can be found in Richard Stewart and Jonathan Wiener, *Reconstructing Climate Policy: Beyond Kyoto* 49–53 (2003).

23. Cf. Adrian Vermeule, Reparations As Rough Justice 15, University of Chicago Law School, John M. Olin Law and Economics Working Paper No. 260; University of Chicago Law School, Public Law and Legal Theory Working Paper No. 105), available at http://www.law.uchicago.edu/law-pdf/law-econ/260.pdf (explaining that rough justice might be the best justification for reparations).

24. Cf. Jacob T. Levy, *The Multiculturalism of Fear* 242 (2000). Levy argues that such people should feel shame about national failures, and not exactly that they have any moral obligations. However, the latter view seems to reflect many people's intuitions.

Chapter 6: Equality and the Case against Per Capita Permits

1. The number of authors arguing for this approach is large. For a small sampling, see, e.g., National Development and Reform Commission, People's Republic of China, China's National Climate Change Programme 58 (June 2007); Anil Agarwal and S. Narain, *Global Warming in an Unequal World: A Case of Environmental Colonialism* (1991); Tom Athanasiou and Paul Baer, *Dead Heat: Global Justice and Global Warming* (2002); Ann Kizig and Daniel Kammen, National Trajectories of Carbon Emissions; Analysis of Proposals to Foster the Transition to Low-Carbon Economies, 8 *Global Environmental Change* 183 (1998); Juan-Carlos Altamirano-Cabrera and Michael Finus, Permit Trading and Stability of International Climate Agreements, 9 *Journal of Applied Economics* 19 (2006); A. D. Sagar, Wealth, Responsibility, and Equity: Exploring an Allocation Framework for Global GHG Emissions, 45 *Climatic Change* 511 (2000); Peter Singer, *One World: The Ethics of Globalization* 35 (2002); Hermann E. Ott and Wolfgang Sachs, The Ethics of International Emissions Trading, in *Ethics, Equity, and International Negotiations on Climate Change* 159 (Luiz Pinguelli-Rosa and Mohan Munasinghe, eds., 2002); Malik Amin Aslam, Equal Per Capita Entitlements: A Key to Global Participation on Climate Change? in *Building on the Kyoto Protocol: Options for Protecting the Climate* 127 (Kevin A. Baumert, ed., 2002); Donald Brown, *American Heat: Ethical Problems with the United States' Response to Global Warming* 214 (2002).

2. See, e.g., National Development and Reform Commission, supra n.1; Frankel, supra n.1; Altamirano-Cabrera and Finus, supra n.1.

3. Michael Grubb, et al., *Sharing the Burden, in Confronting Climate Change: Risks, Implications and Responses* 270 (Irving M. Mintzer, ed., 1992).

4. Roberts and Parks, supra n.1, at 144.

5. See, e.g., Grubb, supra n.3; Singer, supra n.1.

6. The Law of the Sea Convention provides that such resources be divided "equitably"; however, that term has multiple meanings and is left undefined. See United Nations Convention on the Law of the Sea, Art. 140.

7. Under the status quo approach, the United States would be allocated about 20 percent of the permits. Under a per capita approach, the United States

would be allocated about 5 percent of the permits (the U.S. share of the global population). Assuming that the price of a permit is $25 per metric ton of CO_2 (the current price in the EU market) and enough permits are supplied to permit the output rate of 30 billion metric tons per year (roughly the current global rate), then moving from the status quo approach (6 billion tons) to the per capita approach (1.5 billion tons) would cost the United States about $112.5 billion per year. These are back-of-the-envelope calculations intended to give a rough sense of the magnitude involved, and should be taken with many grains of salt.

8. The number may be too high because nations can reduce their per capita emissions and hence, the net outflow. On the other hand, if per capita emissions increase, say because of economic development, the number will be too high.

9. *The Garnaut Climate Change Review* supports this approach and cites numerous governments and international commissions that do so as well. See Ross Garnaut, *The Garnaut Climate Change Review* 203 (2008).

10. We put to one side some prominent puzzles about the relationship between happiness and income. See Robert Frank, *Luxury Fever* (1992); Richard Layard, *Happiness* (2004).

11. Matthew Adler and Eric A. Posner, *New Foundations for Cost-Benefit Analysis* (2005).

12. These effects are addressed in the law and economics literature on legal transitions. See, e.g., Louis Kaplow, An Economic Analysis of Legal Transitions, 99 *Harvard Law Review* 509 (1986). This literature focuses on domestic law, where it is clearer than in the international context that a government that adopts certain policies or practices toward legal transitions—compensating or grandfathering those injured by the transition, for example—will affect the incentives of people to anticipate legal change. We extend this argument to the international setting; there is no reason to think that the differences in settings should affect the analysis. The transitions literature ignores what we have called "ex post efficiency," instead assuming that whatever legal change is introduced is dictated by efficiency. The environmental literature, by contrast, focuses on ex post efficiency (for example, the choice between permits systems and taxes) and generally although not always ignores ex ante issues. For a discussion, see Jonathan R. Nash, Allocation and Uncertainty: Strategic Responses to Environmental Grandfathering, unpublished manuscript 18–22 (2008).

13. In fact, development economists have gone so far as to identify a "resource curse": countries with valuable natural resources often do worse than those that lack them. See Richard M. Auty, *Sustaining Development in Mineral Economies: The Resource Curse Thesis* (1993).

14. David Dollar and Victoria Levin, The Increasing Selectivity of Foreign Aid, 1984–2003, 34 *World Development* 2034 (2006).

15. Hence, the scholarly support for banking systems under which any future climate treaty would reward states that make abatements efforts prior to treaty ratification. See, e.g., Ann P. Kinzig and Daniel M. Kammen, National

Trajectories of Carbon Emissions: Analysis of Proposals to Foster the Transition to Low-Carbon Economies, 8 *Global Environment Change* 183 (1998).

16. See Kathryn Harrison and Lisa McIntosh Sundstrom, The Comparative Politics of Climate Change, 7 *Global Environmental Policy* 1 (2007).

17. Grubb, supra n.3, at 318; Hermann E. Ott and Wolfgang Sachs, The Ethics of International Emissions Trading 168, in *Ethics, Equity and International Negotiations on Climate Change* (Luiz Pinguelli-Rosa and Mohan Munasinghe, eds., 2002).

18. See supra n.6.

19. See E. P. Thomson, *Customs in Common: Studies in Traditional Popular Culture* (1993).

20. We reject this argument for welfarist reasons: it is better to give people incentives to anticipate exogenous changes that necessitate government action than to insure them against losses resulting from that action. See Kaplow, supra n.12.

21. UN Human Rights Council, Report of the Working Group on the Right to Development on its Eighth Session, UN Doc. A/HRC/4/47, ¶ 19 (2007).

22. Ibid. at ¶ 18.

23. See, e.g., supra n.6.

24. These acknowledgments can be found, in vague terms, in such documents as UN General Assembly, Declaration of the Right to Development, Res. 41/28 (1986), and UN General Assembly, Vienna Declaration and Programme of Action, A/CONF.157/23 (1993).

25. On such agreements in general, see Cass R. Sunstein, *Legal Reasoning and Political Conflict* (1996).

26. The best evidence for this proposition is the pattern of foreign aid. Poor countries, understandably, do not provide foreign aid, but middle-income countries also do not seem to feel that they have a responsibility to help people living in poorer countries. Rich countries provide foreign aid but are not generous, and scholars have shown that much (but not all) foreign aid can be traced to specific strategic interests. See, e.g., Alberto Alesina and David Dollar, Who Gives Foreign Aid to Whom and Why? 5 *Journal of Economic Growth* 33, 55–6 (2000).

27. See OECD statistics, available at http://www.oecd.org/dataoecd/42/30/40039096.gif.

28. For a sophisticated demonstration of this problem, see Juan-Carlos Altamirano-Cabrera and Michael Finus, Permit Trading and Stability of International Climate Agreements, 9 *Journal of Applied Economics* 19 (2006). They argue that equitable schemes are more likely to be unstable than "pragmatic" schemes that take account of relative economic power.

Chapter 7: Future Generations

1. The literature is vast. Articles are collected in *Discounting for Time and Risk in Energy Policy* (1982); *Discounting and Intergenerational Equity* (P. R. Portney

and J. P. Weyant, eds., 1999); and Symposium, Intergenerational Equity and Discounting, 74 *University of Chicago Law Review* (2007). The discount rate used in the Stern Review generated a number of comments. For a sample, see Martin Weitzman, A Review of the Stern Review, 45 *Journal of Economic Literature* 703 (2007); and William D. Nordhaus, A Review of the Stern Review on the Economics of Climate Change, 45 *Journal of Economic Literature* 686 (2007).

2. Nicholas Stern, *The Economics of Climate Change* (2007); William Nordhaus, *A Question of Balance* (2008).

3. Nordhaus, supra n.2; Chris Hope, The Marginal Impact of CO_2 from PAGE2002: An Integrated Assessment Model Incorporating the IPCC's Five Reasons for Concern, 6 *Integrated Assessment* 19 (2006).

4. Working Group II: Impacts Adaptation, and Vulnerability, IPCC Fourth Assessment Report 823, *Climate Change 2007.*

5. Frank Ramsey, A Mathematical Theory of Savings, 38 *Economic Journal* 543 (1928).

6. Roy Harrod, *Towards a Dynamic Economics* (1948).

7. Tjalling C. Koopmans, On the Concept of Optimal Economic Growth, 28 *Pontificae Academiae Scientiarum Scripta Varia* 225 (1965).

8. See supra n.1 for a partial list of references.

9. This number is in the ballpark of current figures. See Cass R. Sunstein, Valuing Life: A Plea for Disaggregation, 54 *Duke Law Journal* 385 (2004) (listing a figure of $6.1 million).

10. This is not to say that the issues are not important. Because small changes in the discount rate can have large effects on project valuation, it is very important to get these issues right. They are, however, not central to our discussion.

11. Martin L. Weitzman, Why the Far-Distant Future Should Be Discounted at Its Lowest Possible Rate, 36 *Journal of Environmental Economics and Management* 201 (1998).

12. Thomas Sterner and U. Martin Persson, An Even Sterner Review: Introducing Relative Prices into the Discounting Debate, Resources for the Future Discussion Paper. Washington, DC: Resources for the Future (2007).

13. For summaries of this approach to discounting, see Nicholas Stern, *The Economics of Climate Change* (2007); and Geoffrey Heal, Intertemporal Welfare Economics and the Environment, in *The Handbook of Environmental Economics* 1106 (K-G. Maler and J. R. Vincent, eds., 2005).

14. This simplification is simply astonishing. Stern, for example, severely criticizes the positivists for requiring all kinds of specialized assumptions for the private rate of return to equal the social rate of return, but then imposes specialized functional forms. Stern, *The Economics of Climate Change.* Although the use of this functional form has a long history in public economics, it remains a specialized assumption. See Anthony B. Atkinson, Measurement of Inequality, 2 *Journal of Economic Theory* 244 (1970).

15. Stern, supra n.2.

16. We use $95 rather than $100 because the newly discovered environmental harm means that we (all generations together) are not as well off as we thought. It is as if we lost money. It is likely that all generations will need to share in this loss. The number is only illustrative and we take no position on whether it should be more or less than $100 because of the damages from climate change.

17. This is a simplification of the idea of distribution-neutral investment choice discussed in detail in Louis Kaplow, Discounting Dollars, Discounting Lives: Intergenerational Justice and Efficiency, 74 *University of Chicago Law Review* 79 (2007); Dexter Samida and David Weisbach, Paretian Intergenerational Discounting, 74 *University of Chicago Law Review* 145 (2007).

18. An alternative, slightly more controversial way to make this point is that the ethicists observe that the private rate of return is not equal to the social rate of return and suggest that the government can fill this gap. For example, if the private market rejects a project because the rate of return is only, say 5 percent when it demands a 5.5 percent return, the government should engage in the project if the social rate of return is lower, such as the 1.4 percent used by Stern. Given large differences in the private rate of return and the social rate of return, the government would be engaging in a vastly greater number of projects than any democratic government currently does. There are likely to be good reasons for restricting the scope of government projects, however. Therefore, the ethicists' arguments for a very low social discount rate are incomplete. Recommendations about government projects using a low social discount rate need to be combined with these reasons for restricting government projects. The models run by the ethicists and the resulting recommendations, however, never include these exogenous restrictions.

19. A more subtle and less powerful objection is that even if overall savings rates stay the same, interest rates may change with a change in projects. For example, if we keep our legacy to the future at $100 but change the mix of projects that make up this $100, market rates of return may change. This concern seems second order and interest rates could go up as well as down. See Kaplow, supra n.17 for a discussion.

20. For a review, see Kent Smetters, Ricardian Equivalence: Long-Run Leviathan, 73 *Journal of Public Economics* 395 (1999).

21. Robert C. Lind, A Primer on Major Issues Relating to the Discount Rate for Evaluating National Energy Options, in *Discounting for Time and Risk in Energy Policy* 21–4 (Robert C. Lind, ed., 1982); Robert C. Lind, Analysis for Intergenerational Discounting, in *Discounting and Intergenerational Equity* (Paul R. Portney and John P. Weyant, eds., 1999).

22. Lind, Analysis for Intergenerational Discounting, supra n.21.

23. Richard L. Revesz, Environmental Regulation, Cost-Benefit Analysis, and the Discounting of Human Lives, 99 *Columbia Law Review* 941 (1999).

24. See, e.g., W. Kip Viscusi, *Fatal Tradeoffs* (1994).

25. For discussion, see Cass R. Sunstein, Valuing Lives: A Plea for Disaggregation, 54 *Duke Law Journal* 385 (2004).

26. For a detailed discussion of this point, see Cass Sunstein and Arden Rowell, On Discounting Regulatory Benefits: Risk, Money, and Intergenerational Equity, 74 *Chicago Law Review* 171 (2007).

27. Derek Parfit, *Reasons and Persons* (1984).

28. Viscusi, supra n.24; Sunstein, supra n.25.

Chapter 8: Global Welfare, Global Justice, and Climate Change

1. See, e.g., Martha Nussbaum, *Frontiers of Justice* (2006).

2. See, e.g., Peter Singer, *One World: The Ethics of Globalization* (2002).

3. See James Griffin, *Well-Being: Its Meaning, Measurement, and Moral Importance* (1986).

4. For relevant discussion, see Amartya Sen, *Development as Freedom* (2000).

5. See Nussbaum, supra n.1.

6. See, e.g., Peter Singer, Famine, Affluence, and Morality, 1 *Philosophy and Public Affairs* 229 (1972).

7. Utilitarianism is a particular form of welfarism. Welfarism in general is a belief that the value of a policy depends on the effects of aggregate welfare. Many different forms of aggregation are consistent with welfarism. Utilitarianism is a belief that aggregate welfare should be measured by the sum of the welfare of individuals.

8. Probably the closest precedent was the abolition of the slave trade, which was mainly the work of one nation—Great Britain—which exploited its control of the sea.

9. The two problems will, in practice, be intertwined because nations will claim that they are not free-riding; they will say that the treaty does not make them better off. Because of the difficulties in calculating the costs of emissions reductions and the local benefits of climate change abatement, it will be difficult to refute such claims.

10. H.L.A. Hart, Are There Any Natural Rights? 64 *Philosophical Review* 185 (1955); John Rawls, Legal Obligation and the Duty of Fair Play, in *Law and Philosophy: A Symposium* (Sidney Hook, ed., 1964). Robert Nozick has objected to these arguments. See Robert Nozick, *Anarchy, State, and Utopia* (1974), chapter 5, using as an example, an obligation to contribute to a public address system that you view as having marginal value.

11. Rawls, supra n.10.

Index